Environmental Science and Green Computing

Dr.T. Indumathi

Dr.P. Prabu

Published by

Environmental Science and Green Computing

Copyright © 2018 by Bonfring

ISBN 978-93-86638-62-5

Authors

Dr.T. Indumathi

Dr.P. Prabu

Bonfring

309, 2nd Floor, 5th Street Extension, Gandhipuram,

Coimbatore-641 012.

Tamilnadu, India.

E-mail: info@bonfring.org

Website: www.bonfring.org

Phone: 0422 4213231

Acknowledgement

We sincerely thank the almighty God, who is the source of life and strength of knowledge and wisdom.

We would like to express our gratitude to the many people who saw us through this book; to all those who provided support, talked things over, read, wrote, offered comments, allowed us to quote their remarks and assisted in the editing, proofreading and design.

We also wish to thank the management of Sri Krishna College of Engineering and Technology, PSG College of Technology, Coimbatore and Christ University, Bangalore for the academic support and the facilities provided to carry out our text book work.

We would like to thank our family for standing beside us throughout our career and writing this book. Their motivation continued us to improve our knowledge towards the development of our career.

We also thank all our friends and colleagues for their constant source of encouragement and help in various stages of writing this book. Our sincere thanks to our Publisher, for their help and co-operation in publishing this book.

Environmental Science and Green Computing

UNIT I: NATURAL RESOURCES

Introduction-Forest resources: Use and abuse, case study-Major activities in forest-Water resources- Use and overutilization water, dams-benefits and problems–Mineral resources-Use and exploitation, environmental effects of mining- case study–Food resources- World food problems, case study – Energy resources -Renewable and non-renewable energy sources – Land resources- Soil erosion and desertification – Role of an individual in conservation of natural resources.

UNIT II: ECOSYSTEMS AND BIODIVERSITY

Concept of an ecosystem – structure and function of an ecosystem–producers, consumers and decomposers - Carbon cycle and Nitrogen cycle–energy flow in the ecosystem–ecological Pyramid – Introduction, types, characteristic features, structure and function of the forest ecosystem, grassland ecosystem, desert ecosystem and aquatic ecosystems.

Introduction to biodiversity definition: genetic, species and ecosystem diversity – value of biodiversity-India as a nation of mega-diversity – hot-spots of biodiversity – threats to biodiversity– Endangered and endemic species – conservation of biodiversity: In-situ and ex-situ conservation of biodiversity.

Field study of common plants, insects, birds. Field study of simple ecosystems – pond, river, hill, slopes.

UNIT III: ENVIRONMENTAL POLLUTION

Definition – causes, effects and control measures of: (a) Air pollution-Acid rain-Green house effect-Global warming- Ozone layer depletion – case study- Bhopal gas tragedy (b) Water pollution (c) Soil pollution - Solid waste management: causes, effects and control measures of municipal solid wastes – (d) Noise pollution (e) Thermal pollution (f) Nuclear hazards-case study-Chernobyl nuclear disaster-Role of an individual in prevention of pollution –Field study of local polluted site – Urban / Rural / Industrial / Agricultural.

UNIT IV: SOCIAL ISSUES AND THE ENVIRONMENT

Sustainable development-water conservation, rain water harvesting, watershed management – resettlement and rehabilitation of people; its problems and concerns, case study – Wasteland reclamation - Environmental ethics: 12 Principles of green chemistry-Scheme of labeling of environmental friendly products (Ecomark) – Emission standards – ISO 14001 standard.

UNIT V: HUMAN POPULATION AND THE ENVIRONMENT

Population growth, variation among nations – population explosion – family welfare programme – environment and human health – human rights– value education – HIV / AIDS – woman and child welfare –Role of information technology in environment and human health– Case study.

Self-Study: Air pollution in industrial areas, water pollution studies on river, pond, lake and other water bodies.

UNIT VI: GREEN COMPUTING AND E-WASTAGE

Introduction-Importance of Green Computing-Greening of IT-Understanding the Importance of Green IT in Business-Can IT make your company Green-Green Internet of Things-Green Devices-E-wastage-Process of E-waste Recycling-Green Engineering.

Environmental Science

Common to all branches

Introduction to Environmental Science

Definition

The **environment** is derived from the French word **Environ** which means to encircle or surround. The environment is the sum total of water, air, and land, inter-relationships among themselves and also with the human beings, other living organisms and property. The above definition is given in Environment Act, 19860 clearly indicates that environment includes all the physical and biological surroundings and their interactions.

Scope and Importance

The scope of environmental science is broad. Some of the aspects are

- Studying the interrelationships among biotic and abiotic components for sustainable human ecosystem
- Carrying out impact analysis and environmental auditing for the further catastrophic activities
- Developing and curbing the pollution from existing and new industries
- Stopping the use of biological and nuclear weapons for destruction of human race
- Managing the unpredictable disasters and so on.

In recent years, the scope has also emerged in the form of career options such as

- R & D
- Green advocacy
- Green marketing
- Green media
- Environment consultancy

Need for Public Awareness

Environmental issues received international attention about 35 years back in Stockholm conference, held on 5th June 1972. The United Nations Conference on Environment and Development held in Rio de Janerio in 1992 popularly known as Earth summit followed by the world summit on sustainable Development in 2002, have highlighted key issues of global environmental concern such as:

- Public awareness is essential to help understand pros and cons of environmental problems.
- Environmental pollution cannot be removed by laws alone.

- The proper implementation especially public participation is one of the important aspects.
- Public participation is possible only when the public is aware of the ecological and environmental issues.
- A drive by the government to ban the littering of polyethene cannot be successful until the public understands the environmental complications of the same.
- The public has to be educated about the fact that if we are degrading our environment we are actually harming ourselves.

Risk and Hazards

Hazard is a substance which creates illness or harmful effects

Hazard = f(risk X exposure X vulnerability X response)

Types of Hazard

1. Physical hazards – Ex: cold , heat , noise
2. Chemical hazards – Ex : toxic vapours , flammable gases
3. Biological hazards – Ex: Bacteria , virus

<table>
<tr><th>Unit</th><th>Contents</th><th>Page No</th></tr>
</table>

UNIT I

NATURAL RESOURCES

1.1.   Introduction

Life on this planet earth depends on upon a large number of things and services provided by nature, which is known as Natural resources. Thus water, air, soil, minerals, coal, forests, crops, and wildlife are all examples of natural resources. [Valuable resources of the environment are termed as natural resources].

Ex: Soil, water, air, minerals.

The natural resources are of two kinds:

Renewable Resources

Renewable resources which are in exhaustive and can be regenerated within a given span of time e.g. forests, wildlife, wind energy, biomass energy, tidal energy, hydro power etc. Solar energy is also a renewable form of energy as it is an inexhaustible source of energy.

[In exhaustive and can be regenerated within a given span of time]

Ex: Wind energy, Tidal energy, Hydropower.

Non-Renewable Resources

Non-renewable resources which cannot be regenerated.

Ex: Coal, petroleum, minerals

Once we exhaust these reserves, the same cannot be replenished.

[Exhaustive and cannot be regenerated.]

It is the responsibility of all the individuals to protect these resources in a judicial manner.

Some of the major natural resources are:

1. Forest resources.
2. Water resources.
3. Mineral resources.
4. Food resources.
5. Energy resources.
6. Land resources.

1.2. Forest Resources

World Food Day – October 16th

Forests are the important natural resources on this earth. About 1/3 land area in the world is covered with forest. Latest reports indicate that the current deforestation level is to a great extent in Brazil.

Uses of Forests

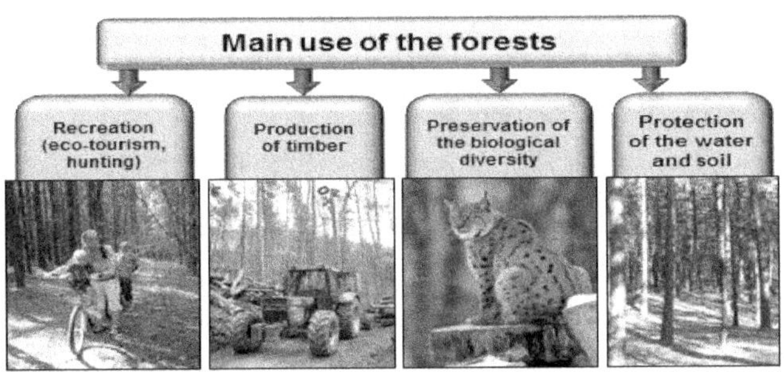

Commercial Uses

- Man depends heavily on a larger number of plant and animal products from forests for his daily needs.
- The chief product that forests supply is wood, which is used as fuel, the raw material for various industries as pulp, paper, newsprint, board, timber for furniture items, other uses as in packing articles, matches, sports goods etc.
- Indian forests also supply minor products like gums, resins, dyes, tannins, fibers, etc.
- Many of the plants are utilized in preparing medicines and drugs; Total worth of which is estimated to be more than $300 billion per year.
- Many forests lands are used for mining, agriculture, grazing, and recreation and for the development of dams.

Ecological Uses

The ecological services provided by our forests may be summed up as follows:

Production of Oxygen: The main greenhouse gas carbon dioxide is absorbed by the forests as a raw material for photosynthesis. Thus, forests act as a sink for carbon dioxide thereby reducing the problem of global warming caused by greenhouse gas CO_2

Wildlife habitat: Forests are the homes of millions of wild animals and plants. About 7 million species are found in the tropical forests alone.

Regulation of hydrological Cycle: Forested watersheds act like giant sponges, absorbing the rainfall, slowing down the runoff. They control climate through transpiration of water and seed clouding.

Soil Conservation: Forests bind the soil particles tightly in their roots and prevent soil erosion. They also act as wind breakers.

Pollution moderators: Forests can absorb many toxic gases and can help in keeping the air pure and in preventing noise pollution.

Over Exploitation of Forests

Exploitation of forests takes place due to the deficiency of human needs

- Man depends heavily on forests for food, medicine, shelter, wood and fuel.
- With growing civilization the demands for raw material like timber, pulp, minerals, fuel wood etc. shot up resulting in large scale logging, mining, road-building and clearing of forests.
- The international timber trade alone is worth over US $ 40 billion per year.
- The devasting effects of deforestation in India include soil, water, and wind erosion, estimated to cost over 16,400 crores every year.
- Big hydropower projects result in large scale destruction of forests.

Deforestation

Deforestation means the destruction of forests.

- The total forests area of the world in 1900 was estimated to be 7,000 million hectares which were reduced to 2890 million ha in 1975 fell down to just 2,300 million ha by 2000.
- Deforestation rate is relatively less in temperature countries, but it is very alarming in tropical countries.
- Deforestation is a continuous process in India where about 1.3 hectares of forest land has been lost.
- The per capita availability of forest in India is 0.08 hectares per person which is much lower than the world average of 0.8 hectares.
- The presence of waste land is a sign of deforestation in India.

Causes of Deforestation

Major causes of deforestation are listed below:

- Development projects.
- Shifting cultivation.
- Fuel requirements.

- Construction of dams.
- Growing food needs.
- Raw materials for industrial usage.

Consequences of Deforestation

Some of the effects of deforestation are listed below:

a) Effect on climate.
- Global warming.
- Less rainfall.
- Hot climate and others.

b) Effect on biodiversity.
- Lose of medicinal plants.
- Lose of timber fuels wood and others.

c) Effect on resources.
- Lose of land resource.
- Lose of soil fertility.
- Soil erosion.
- Drastic changes in biogeochemical cycles.

d) Effect on economy.
- Increase in medicinal values.
- Demand for industrial products and others.

Case Studies

Desertification in Hilly Regions of the Himalayas

Desertification in the Himalayas, involving clearance of natural forests and plantation of monocultures like Pinus Roxburgh, Eucalyptus camadulensis etc., have upset the ecosystem by changing the various soil and biological properties. The area is invaded by exotic weeds. These areas are not able to recover and are losing their fertility.

Timber Extraction

Logging for valuable timber such as teak and mahogany not only involves a few large trees per hectare but about a dozen more trees since they are strongly interlocked with each other by vines etc. Also, road construction for making an approach to the trees causes further damage to the forests. In India, firewood demand would continue to rise in future mostly consumed in rural areas, where alternative sources of energy, are yet to reach.

Mining

Mining is the process of removing deposits of ores from substantially very well below the ground level. Mining is carried out to remove several minerals including coal. These mineral deposits invariably found in the forest region, and any operation of mining will naturally affect the forests.

Mining from shallow deposits is done by surface mining while that from deep deposits is done by sub-surface mining. More than 80,000 ha of land of the country is presently under the stress of mining activities.

Effects of Mining Resources

Mining operation requires removal of vegetation along with underlying soil mantle and overlying rock masses. This results in the destruction of the landscape in the area.

- The Large scale of deforestation has been reported in Mussoorie and Dehradun valley due to mining of various areas.
- Indiscriminate mining in Goa since 1961 has destroyed more than 50,000 ha of forest land.
- Mining of radioactive mineral in Kerala, Tamilnadu and Karnataka are posing similar threats of deforestation.

Dams and their Effects on Forests and Tribal People

Big dams and river valley projects have multi-purpose uses and have been referred to as "Temples of modern India". India has more than 1550 large dams, the maximum being in the state of Maharashtra (more than 600) followed by Gujarat (more than 250) and Madhya Pradesh (130).The highest one is Tehri dam, on river Bhagirathi in Uttaranchal and the largest in terms of capacity is Bhakra dam on river Sutlej.

Effects on Tribal People

- The greatest social cost of the big dam is the widespread displacement of local people.
- It is estimated that the number of people affected directly or indirectly by all big irrigation projects in India over the past 50 years can be as high as 20 million.
- The Hirakud dam, one of the largest dams executed in the fifties, has displaced more than 20,000 people residing in 250 villages.

Effects on Forests

- Thousands of hectares of forests have been cleared for executing river valley projects which breaks the natural ecological balance of the region. Floods, landslides become more prevalent in such areas.

For example

- The Narmada Sagar project alone has submerged 3.5 lakh hectares of best forest comprising of rich teak and bamboo forests.
- The Tehri dam submerged 1000 hectares of forest affecting about 430 species of plants according to the survey carried out by the botanical survey of India.

1.3. Water Resources

Water is an indispensable natural resource which covers around 97% on the earth.

Uses of Water

Due to its unique properties, water is of multiple uses for all living organisms.

- Water is absolutely essential for life.
- Most of the life processes take place in water contained in the body.
- Uptake of nutrients, their distribution in the body, regulation of temperature, and removal of wastes are all mediated through water.
- Human beings depend on water for almost every developmental activity.
- Water is used for drinking, irrigation, and transportation, washing and waste disposal for industries and used as a coolant for thermal power plants.
- Water shaped the earth's surface and regulates our climate.

Overutilization of Surface and Ground Water

- With increasing human population and rapid development, the world water withdrawal demands have increased many folds and a large proportion of the water withdrawn is polluted due to anthropogenic activities.
- Out of the total water reserves of the world, about 97% is salt water and only 3% is fresh water.
- Even this small fraction of fresh water is not available to us as most of it is locked up in polar ice caps and just 0.003% is readily available to us in the form of groundwater and surface water.

Ground Water

Ground water was considered to be the purest form. But nowadays even the ground water has been found to be contaminated by leachates from sanitary landfills.

Aquifer

A layer of sediment rock that is highly permeable containing water

Confined Aquifer

Sandwiched between two impermeable layers of rock and recharged only in those areas where the aquifer intersects the land surface.

Unconfined Aquifers

Overlain by permeable earth materials and are recharged by the water seeping down in the form of rainfall and snow melt.

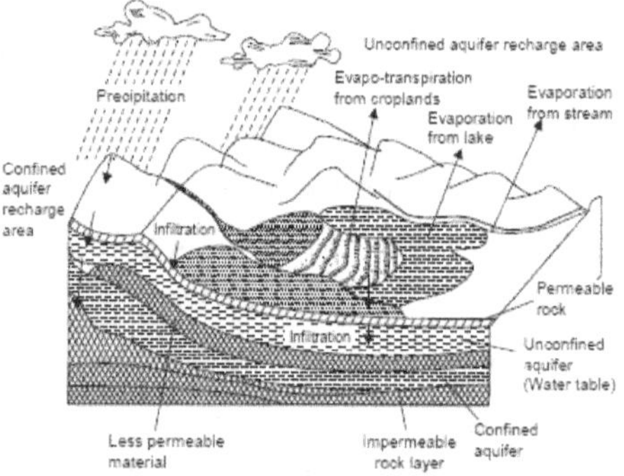

Fig. The groundwater system. An unconfined aquifer (water table) is formed when water collects over a rock or compact clay. A confined aquifer is formed sand switched between two layers having very low permeability.

Effects of Over Exploitation of Ground Water

Subsidence: When groundwater withdrawal is more than its recharge rate, the sediments in the aquifer (a layer of rock that is highly permeable and contains water) get compacted, a phenomenon known as ground subsidence. It results in the sinking of the overlying land surface. Due to this structural damage in buildings, a fracture in pipes etc. occurs.

Lowering of water table: Mining of groundwater is done extensively for irrigating crop fields. However, excessive mining would cause lowering of water table.

Water logging: When excessive irrigation is done with brackish water it raises the water table gradually leading to water-logging and salinity problems.

Surface Water

The water which assumes to be in the form of lakes, ponds, rivers is known as surface water.

Floods and Drought

- Heavy rainfall often causes floods in the low-lying coastal areas.
- A prolonged downpour can also cause the overflowing of lakes and rivers resulting into floods.
- When annual rainfall is below normal and less than evaporation, drought conditions are created.

Causes of Flood and Drought

- Deforestation, overgrazing, mining, rapid industrialization, global warming etc., have contributed largely to a sharp rise in the incidence of floods.
- Deforestation leads to desertification and drought too. When the trees are cut, the soil is subject to erosion by heavy rains, winds, and the sun.
- The removal of a thin top layer of soil takes away the nutrients and the soil becomes useless.
- The eroded soils exhibit drought tendency.

Conflicts over Water

The indispensability of water and its unequal distribution has often led to inter-state or international disputes. Issues related to sharing of river water have been largely affecting our farmers and also shaking our governments. Many countries are engaged in bitter rivalries over this precious resource.

For instance

- Argentina and Brazil, dispute each other's claims to the La Plata river.

- India and Pakistan fight over the rights to water from the Indus.

- Mexico and the USA have come in conflict over the Colorado river.

- India and Bangladesh are fighting for Brahmaputra river.

- Iran and Iraq contest for the water from Shatt-Al- Arab River.

- Within India, water conflicts are still being continues between the states.

- Sharing of Krishna water between Karnataka and Andhra Pradesh.

- Sharing of Siruvani water between Tamilnadu and Kerala, and others.

- Sharing of Cauvery between Karnataka and Tamilnadu.

- On June 2,1990, the Cauvery Water dispute Tribunal was set up which through an interim award directed Karnataka to ensure that 205 TMCF of water was made available in Tamil Nadu's Mettur dam every year, till a settlement was reached.

- In 1991-1992 due to good monsoon, there was no dispute. In 1995, the situation turned into a crisis due to delayed rains and an expert Committee was set up to look into the matter which found that there was a complex cropping pattern in Cauvery basin.

- Samba paddy in winter, Kuravai paddy in summer and some cash crops demanded intensive water; thus aggravating the water crisis.

- Proper selection of crop varieties, optimum use of water, better rationing are suggested as some measures to solve the problem.

Big-Dams –Benefits and Problems

Benefits

- River valley projects with big dams play a key role in the development process due to their multiple uses.

- These dams aim at providing employment for tribal people and raising the standard and quality of life.

- Dams can help in checking floods and generate electricity and reduce water and power shortage, provide irrigation water to lower areas, provide drinking water in remote areas and promote navigation, fishery etc.

Problems

The impacts of big dams can be upstream as well as downstream levels. The upstream problems include the following:

- Displacement of tribal people.
- Lose of forests, flora, and fauna.
- Changes in fisheries.
- Siltation and sedimentation of reservoirs.
- Lose of non-forest land.
- Stagnation and water logging near reservoir.
- Breeding vectors and spread of vector –borne diseases.
- Reservoir induces seismicity causing earthquakes.
- Microclimatic changes.
- Growth of aquatic weeds.

The Downstream Problems Include the Following

- Water logging and salinity due to over irrigation.
- Microclimatic changes.
- Reduced water flow and silt deposition in river.
- Flash foods.
- Salt water intrusion at river mouth.
- Lose of land fertility.
- The outbreak of vector-borne diseases likes malaria.

1.4.   Mineral Resources

Minerals are the naturally occurring inorganic, crystalline solids having definite chemical composition and characteristic physical properties. Most of the rocks are composed of minerals such as quartz, feldspar, biotite.

Gypsum Desert rose

Pink Cubic halite crystals

Diamond (NaCl; halide class)

Uses of Minerals

Mineral is an element or inorganic compound that occurs naturally. The main uses of minerals are as follows:

- Development of industrial plants and machinery.
- Generation of energy e.g. coal, lignite, uranium.
- Construction, housing, settlements.
- Defense equipment- weapons, settlement.
- Transportation means.
- Communication-telephone wires, cables, electronic devices.
- Medical system- particularly in Ayurvedic System.
- Formation of alloys for various purposes.
- Agriculture- as fertilizers, seed dressings, and fungicides.
- Jewellery- eg. Gold, silver, platinum, diamond.

Major reserves and important uses of some of the metals.

Metals	Major world reserves	Major uses
Aluminium	Australia, Jamaica	Packing food items, transportation, utensils, electronics
Chromium	CIS(The commonwealth of Independent states), South Africa	For making high strength steel alloys, in textiles and tanning industries
Copper	U.S.A, Canada, CIS	Electronic and electrical goods, building, construction, vessels
Iron	CIS, Canada, U.S.A	Heavy machinery, steel production transportation means.
Manganese	South Africa, CIS	For making high strength heat resistant steel alloys
Platinum	South Africa, CIS	Use in automobiles, catalytic converters, electronics, medical uses.
Gold	South Africa, CIS, Canada	Ornaments, medical use, electronic use, in aerospace
Silver	Canada, South Africa	Photography, electronic jewellery.
Nickel	CIS, Canada	Chemical industry, steel alloys

Major uses of some of the non-metallic minerals.

Non-metal mineral	Major uses
Silicate minerals	Sand and gravel for construction, bricks, paving etc.
Limestone	Used for concrete, building stone, used in agriculture for neutralizing acid soils, used in cement industry
Gypsum	Used in plaster wallboard, in agriculture
Potash, phosphorite	Used as fertilizers
Sulphur pyrites	Used in medicine, car battery, industry

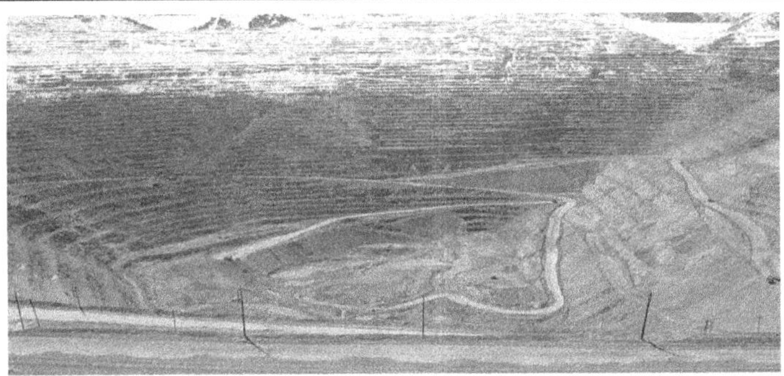

Minerals are sometimes classified as,

Critical Minerals

Essential for the economy of the nation.

Ex: Iron, aluminum.

Strategic Minerals

Required for the defence of a country.

Ex: Mn, Co, Ni.

Environmental Impacts of Mineral Extraction

Major mines which are known for causing severe problems are given below:

- Jaduguda Uranium Mine, Jharkhand- exposing local people to radioactive hazards.
- Jharia coal mines, Jharkhand- underground fire leading to land subsidence and forced displacement of people.

- Sukinda chromite mines, Orissa- Seeping of hexavalent chromium into river posing a serious health hazard, Cr^{6+} being highly toxic and carcinogenic.
- Kudremukh iron ore mine, Karnataka- causing river pollution and a threat to biodiversity.
- East coast Bauxite mine, Orissa-Land encroachment and issue of rehabilitation unsettled.
- North-Eastern Coal Fields, Assam-Very high sulphur contamination of groundwater.

Impacts of mining: Mining is done to extract minerals from deep deposits in the soil. Environmental damages caused by mining activities are as follows:

Devegetation and defacing of lands: Mining requires removal of vegetation along with underlying soil mantle and overlying rock masses. This results in the destruction of the landscape in the area.

Subsidence of land: Subsidence of mining areas results in tilting of buildings, cracks in houses, buckling of roads, bending of rail tracks and leaking of gas from cracked pipelines leading to serious disasters.

Groundwater contamination: Mining pollutes the groundwater. Sulphur, usually present as an impurity in many ores is known to get converted into sulphuric acid through microbial action, thereby making the water acidic.

Surface water pollution: The acid mine drainage often contaminates the nearby streams and lakes. The acidic water, radioactive substances like uranium, heavy metals also contaminate the water bodies and kill aquatic animals.

Air pollution: In order to separate and purify the metal from other impurities in the ore, smelting is done which emits enormous quantities of air pollutants. Oxides of sulphur, arsenic, cadmium and lead etc. shoot up in the atmosphere near the smelters and the public suffers from several health problems.

Occupational Health Hazards: Miners working in different type of mines suffer from asbestosis, silicosis, black lung disease etc

Remedial Measures

- Adopting eco friendly mining technology.
- Utilization of low-grade ores by using microbial – leaching technique. In this method, the pores are inoculated with the desired strains of bacteria like Thiobacillus ferroxidans, which remove the impurities and leave the pure mineral.

- Re-vegetating mined areas with appropriate plants.
- Gradual restoration of flora.
- Prevention of toxic drainage discharge.

Case studies

Mining and Quarrying in Udaipur

Soapstones, building stone, and dolomite mines spread over 15,000 hectares in Udaipur have caused many adverse impacts on the environment. About 150 tonnes of explosives are used per month in blasting. The Maton mines have badly polluted the Ahar river. The hills around the mines are suffering from acute soil erosion.

The waste water flows towards a big tank of "Bag Dara". Due to the scarcity of water people are compelled to use this effluent for irrigation purpose. The animals like tiger, lion, deer, and birds have disappeared from the mining area.

1.5.   Food Resources

World Food Problems

During the last 50 years, world grain production has increased almost three times. The per capita production is increased by about 50%. At the same time, population growth increased at such a rate in less developed countries. Every 40 million people die of undernourishment and malnutrition.

This means that every year our food problem is killing as many people as were killed by the atomic bomb dropped on Hiroshima during World War II. This statistics emphasize the need to increase our food production, and also to control population growth. It is estimated that 300 million are still undernourished.

Impacts of Overgrazing and Agriculture

Overgrazing

Overgrazing can limit livestock production. Overgrazing occurs when too many animals graze for too long and exceed the carrying capacity of a grassland area.

Impacts of Overgrazing

Land degradation: Overgrazing removes the grass cover. The humus content of the soil is decreased and it leads to poor, dry, compacted soil.

Soil erosion: The soil roots are very good binders of soil. When the grasses are removed, the soil becomes loose and susceptible to the action of wind and water.

Loss of useful species: Due to overgrazing the nutritious species like Cenchrus, panicum etc. is replaced by thorny plants like Parthenium, Xanthium etc. These species do not have a good capacity of binding the soil particles and, therefore, the soil becomes more prone to soil erosion.

Agriculture

Traditional Agriculture and Its Impacts

- Usually involves a small plot.
- Simple tools.
- Naturally available water.
- Organic fertilizer and a mix of crops.

Main Impacts

- Deforestation.
- Soil erosion.
- Depletion of nutrients.

Modern Agriculture and Its Impacts

It makes use of hybrid seeds of selected and single crop variety , high-tech equipment, lots of energy subsidies in the form of fertilizers and pesticides

Main Impacts

Impacts related to high yielding verities (HYV): The uses of HYVs encourage monoculture i.e. the same genotype is grown over vast areas. In the case of an attack by some pathogen, there is the total devastation of the crop by the disease due to exactly uniform conditions, which help in the rapid spread of the disease.

Fertilizer-Related Problems

Micronutrient imbalance: Chemical fertilizers have nitrogen, phosphorus and potassium (N,P,K) which are essential macronutrients. Excessive use of fertilizers causes micronutrient imbalance.

For example, excessive fertilizer use in Punjab and Haryana has caused a deficiency of the micronutrient Zinc in the soils, which is affecting the productivity of the soil.

Nitrate Pollution: Nitrogenous fertilizers applied in the fields often leach deep into the soil and ultimately contaminate the ground water. The nitrates get concentrated in the water and when their concentration exceeds 25 mg/L, they become the cause of a serious health hazard called "Blue Baby Syndrome" or methemoglobinemia. This disease affects the infants to the maximum extent causing even death.

Eutrophication: A large proportion of nitrogen and phosphorus used in crop fields is washed off along with runoff water and reach the water bodies causing over nourishment of the lakes, a process known as Eutrophication. (Eu=more, tropic=nutrition). Due to eutrophication, the lakes get invaded by algal blooms. These algal species grow very fast by rapidly using up the nutrients.

The algal species quickly complete their life cycle and die thereby adding a lot of dead matter. The fishes are also killed and there is a lot of dead matter that starts getting decomposed.

Oxygen is consumed in the process of decomposition and very soon the water gets depleted of dissolved oxygen.

This further affects aquatic fauna and ultimately anaerobic conditions are created where only pathogenic anaerobic bacteria can survive. Thus, due to excessive use of fertilizers in the agricultural fields the lake ecosystem gets degraded.

Pesticide-related problems: Thousands of types of pesticides are used in agriculture. The first generation pesticides include chemicals like sulphur, arsenic, lead or mercury to kill the pests. They have number of side effects as discussed below:

Creating resistance in pests and producing new pests: About 20 species of pests are now known which have become immune to all types of pesticides and are known as "Super pests".

The death of non-target organisms: Many insecticides not only kill the target species but also several non-target species that are useful to us.

Biological magnification: Many of the pesticides are non-biodegradable and keep on accumulating in the food chain, a process called biological magnification. This is very harmful.

Water Logging: Over-irrigation of croplands by farmers for good growth of their crop usually leads to water logging.

Inadequate drainage caused excess water to accumulate underground and gradually forms a continuous column of the water table. Under water-logged conditions, pore spaces in the soil get fully drenched with water and the soil- air gets depleted. The water table rises while the roots of plants do not get adequate air for respiration, Mechanical strength of the soil declines, the crop plants get lodged and crop yield falls.

In Punjab and Haryana, extensive areas have become water-logged due to adequate canal water supply or tube-well water.

Preventing excessive irrigation, subsurface drainage technology and bio-drainage with trees like Eucalyptus are some of the remedial measures to prevent water-logging.

Salinity Problem: At present one-third of the total cultivable land area of the world is affected by salts.

Saline soils are characterized by the accumulation of soluble salts like sodium chloride, sodium sulphate, calcium chloride, magnesium chloride etc. in the soil profile. Their electrical conductivity is more than 4 dS/m.

Sodic soils have carbonates and bicarbonates of sodium, the pH usually exceeds 8.0 and the exchangeable sodium percentage (ESP) is more than 15%.

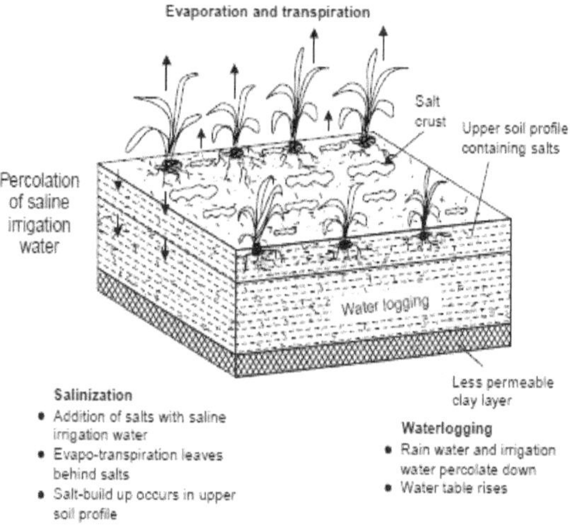

Remedy

1. The most common method for getting rid of salts is to flush them out by applying more good quality water to such soils.
2. Another method is laying an underground network of perforated drainage pipes for flushing out the salts slowly.

Case Studies

Salinity and waterlogging in Punjab, Haryana and Rajasthan:

The first alarming report of salt-affected wasteland formation due to irrigation practices came from Haryana in 1858.Several villages in Panipat and Delhi lying in Western Yamuna Canal were suffering from salinity problems.

The floods of 1947, 1950, 1952, 1954-55 in Punjab resulted in aggravated water logging with serious drainage problems.

Introduction to canal irrigation in 1.3 m ha in Haryana resulted in rising in water table followed by water-logging and salinity in many irrigated areas as a result of fall in crop productivity.

Rajasthan too has suffered badly in this regard following the biggest irrigation project "Indira Gandhi Canal Project"

1.6.　Energy Resources

The first form of energy was probably the fire. Wind and hydropower have also been used for cooking and heating in the last 10000 years. The invention of steam engine has replaced the previous existing factors.

Growing Energy Needs

Development in different sectors relies largely upon energy. Agriculture, industry, mining, transportation, lighting, cooling and heating in buildings all need energy. With the demands of growing population, the world is facing further energy deficit.

In developed countries like U.S.A and Canada, an average person consumes 300 GJ per year. By contrast, an average man in a poor country like Bhutan, Nepal or Ethiopia consumes less than 1 GJ per year. This clearly shows that our lifestyle and standard of living are closely related to energy needs.

Renewable and Non-Renewable Energy Sources

Life on earth depends on upon a large number of things and services provided by nature, which is known as energy resources. Energy Resources are of two kinds.

Renewable resources: which are in exhaustive and can be regenerated within a given span of time eg. Forests, wildlife, wind energy, biomass energy etc. Solar energy is also a renewable form of energy as it is an inexhaustible source of energy.

Non-renewable resources which cannot be regenerated eg. Fossil fuels like coal, petroleum etc. Once we exhaust these reserves, the same cannot be replenished.

Even our renewable resources can become non-renewable if we exploit them to such extent their rate of consumption exceeds their rate of regeneration.

Renewable Energy Resources

Solar Energy

Sun releases an enormous quantity of energy in the form of heat and light. The solar energy received by the near earth space is approximately1.4 kJ/s/m^2 known as solar constant. Now we have several techniques for harnessing solar energy.

Solar heat collectors, solar cells, solar cooker, solar water heater, solar furnace and solar power plant are some important solar energy harvesting devices.

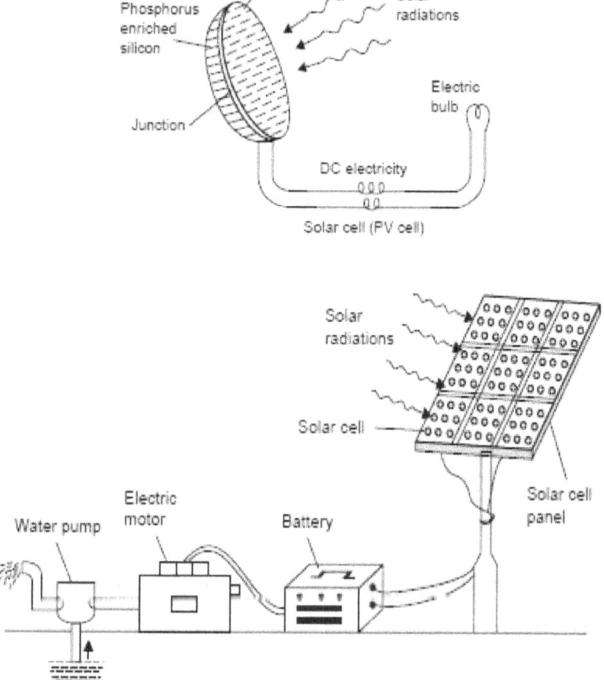

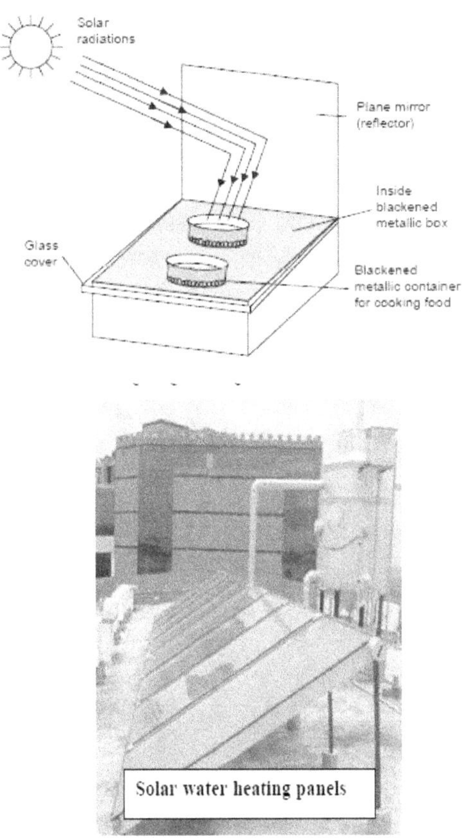

Wind Energy

The high-speed winds have a lot of energy in them as kinetic energy due to their motion.

Wind energy is very useful as it does not cause any air pollution. After the installation cost, the wind energy is very cheap. A large number of windmills together becomes wind farms.

Hydro Power

The water flowing in a river is collected by constructing a big dam where the water is stored and allowed to fall from a height. The blades of turbine located at the bottom of the dam move with the fast moving water which in turn rotates the generator and produces electricity. Hydropower does not cause any pollution. Hydropower projects help in controlling floods, used for irrigation, navigation etc.

Tidal Energy

Ocean tides produced by gravitational forces of sun and moon contain enormous amounts of energy. The tidal energy is harnessed by constructing a tidal barrage. During high tide, the water flows into the reservoir of the barrage and turns the turbine, which in turn produces electricity by rotating the generators.

During low tide, when the sea-level is low, the sea water stored in the barrage reservoir flows out into the sea and again turns the turbines.

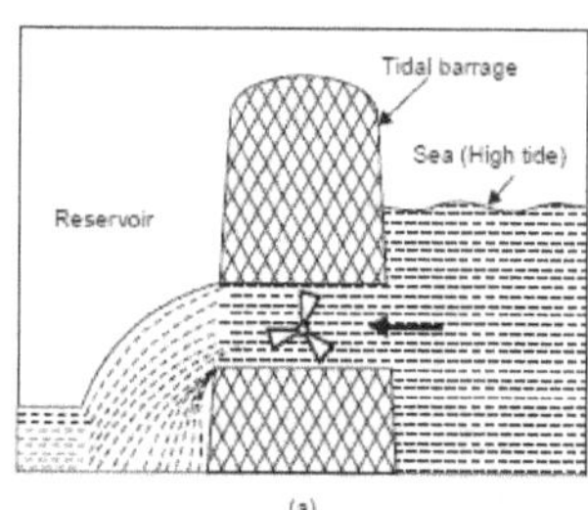

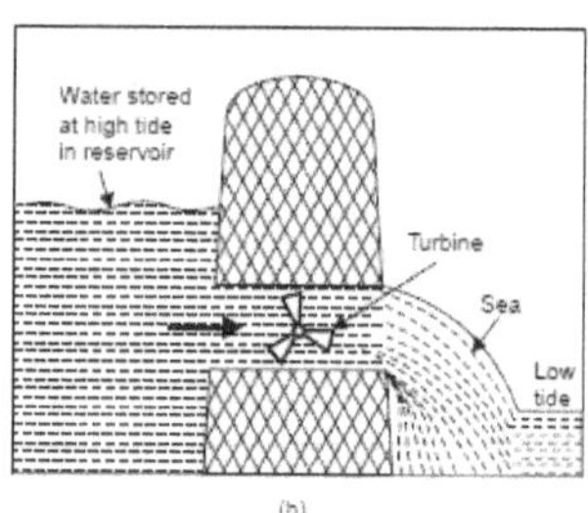

Ocean Thermal Energy (OTE)

The energy available due to the difference in the temperature of water at the surface of the tropical oceans and at deeper levels is called OTE. This energy is used to boil liquid like ammonia. The high-pressure vapours of the liquid formed by boiling are then used to turn the turbine of a generator and produce electricity.

Geothermal Energy

The energy harnessed from hot rocks present inside the earth is called geothermal energy. Sometimes the steam or boiling water underneath the earth does not find any place to come out. We can drill a hole up to the hot rocks and by putting a pipe in it make the steam or hot water gush out through the pipe at high pressure which turns the turbine of a generator to produce electricity.

Biomass Energy

Biomass is the organic matter produced by the plants or animals which include wood, crop, residues, cattle dung agricultural wastes etc. The burning of biogas cause air pollution and produce a lot of ash. It is, therefore, more useful to convert biomass into biogas or biofuels.

Biogas

Biogas is a mixture of methane, carbon dioxide, hydrogen and hydrogen sulphide. Biogas is produced by anaerobic degradation of animal wastes in the presence of water. Anaerobic degradation means the breakdown of organic matter by bacteria in the absence of oxygen. Biogas has many advantages. It is clean, non-polluting and cheap. There is direct supply of gas from the plant and there is no storage problem

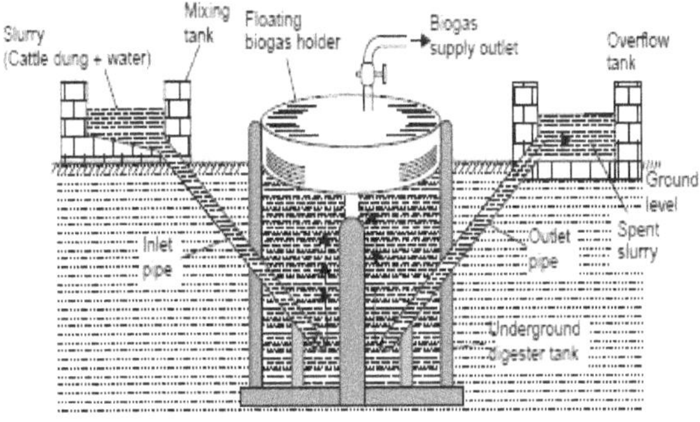

Floating Gas Holder Type Biogas Plant

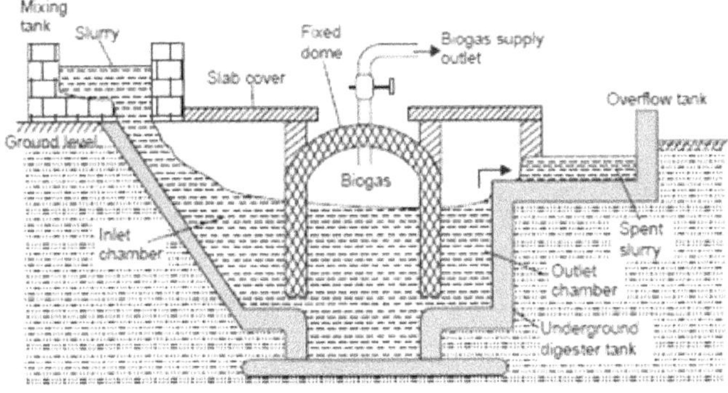

Fixed Dome Type Biogas Plant

Bio Fuels

Biomass can be fermented to alcohols like ethanol and methanol which can be used as fuels. Gasohol is common fuel in Brazil and Zimbabwe for running cars and buses. Methanol is very useful since it burns at a lower temperature than gasoline or diesel. Due to its high calorific value, hydrogen can serve as an excellent fuel. Moreover, it is non-polluting and can be easily produced. Presently H_2 is used in the form of liquid hydrogen as a fuel in spaceships.

Non -Renewable Energy Resources

Coal

Coal was formed 255-250 million years ago in the hot, damp regions of the earth during the carboniferous age. The ancient plants along the banks of rivers were buried after death into the soil and due to the heat and pressures gradually got converted into peat and coal over million years of time. When coal burnt it produces carbon dioxide, which is a greenhouse gas responsible for causing enhanced global warming.

Petroleum

It is the lifeline of the global economy.

Petroleum is a cleaner fuel as compared to coal as it burns completely and leaves no residue. It is also easy to transport and use. Crude petroleum is a complex mixture of alkane hydrocarbons. Hence, it has to be refined by the process of fractional distillation, during which we get a large variety of products namely, petroleum gas, kerosene, petrol, diesel, fuel oil, lubricating oil, paraffin wax etc. The petroleum gas is easily converted to liquid form under pressure as LPG.

Natural Gas

- It is mainly composed of methane with small amounts of propane and ethane.
- It is used as a domestic and industrial fuel in thermal power plants for generating electricity.
- It is used as a source of hydrogen gas in the fertilizer industry and as a source of carbon in tier industry.

Nuclear Energy

- Nuclear energy is known for its high destructive power.
- Nuclear energy can be generated by two types of reactions.

1. **Nuclear fission:** It is the nuclear reaction in which heavy isotopes are split into lighter nuclei on bombardment by neutrons. Fission reaction of U^{235} is given below.

$$_{92}U^{235} + {}_0n^1 \rightarrow {}_{36}Kr^{92} + {}_{56}Ba^{141} + 3\ {}_0n^1 + energy$$

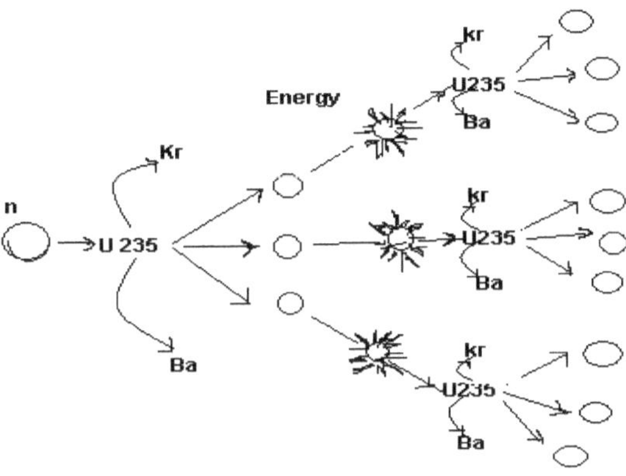

2. **Nuclear fusion:** Here two isotopes of a light element are forced together at extremely high temperatures (1 billion ºC) until they fuse to form a heavier nucleus releasing an enormous amount of energy in the process.

$$_1H^2 + {}_1H^2 \rightarrow {}_3He^2 + {}_0n^1 + energy$$

Nuclear energy has tremendous potential but any leakage from the reactor may cause devastating nuclear pollution. Disposal of the nuclear waste is also a big problem.

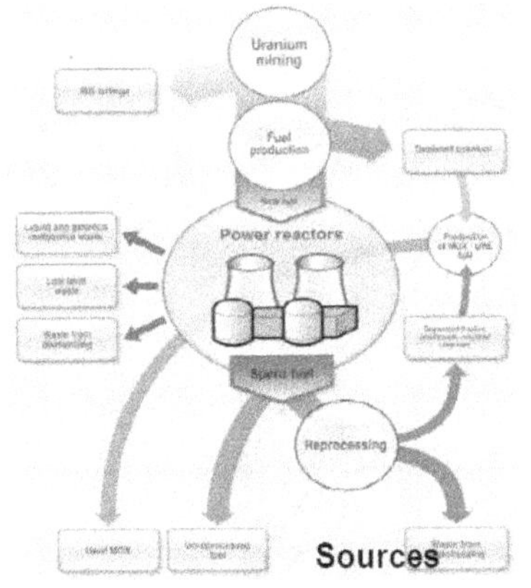

1.7. Land Resources

We depend on upon land for our food, fibre, and fuel wood. About 200-1000 years are needed for the formation of one inch or 2.5 cm soil, depending upon the climate and the soil type. But, when rate of erosion is faster than rate of renewal, then the soil becomes a non-renewable resource

Land Degradation

- With increasing population growth the demands for land for producing food, fibre and fuel wood is also increasing.
- Hence, there is more and more pressure on limited land resources which are getting degraded due to over-exploitation.
- Soil erosion, waterlogging, salinization and contamination of the soil with industrial wastes like fly ash, press mud or heavy metals all cause degradation of land.

Man-Induced Landslides

- Various anthropogenic activities like hydroelectric projects, large dams, reservoirs, construction of roads and railway lines, construction of buildings, mining etc are responsible for the clearing of large forested areas.
- Earlier there were few reports of landslides between Rishikesh and Byasi on Badrinath Highway area. But, after the highway was constructed, 15 landslides occurred in a single year.
- During the construction of roads, mining activities etc. huge portions of fragile mountainous areas are cut or destroyed by dynamite and thrown into adjacent valleys and streams.
- These land masses weaken the already fragile mountain slopes and lead to landslides.
- They also increase the turbidity of various nearby streams, thereby reducing their productivity.

Soil Erosion

Soil erosion is defined as the movement of soil components, especially surface litter and top soil from one place to another. Soil erosion results in the loss of fertility because it is the top soil layer which is fertile.

Soil erosion is basically of two types based upon the cause of erosion:

Normal erosion or geological erosion: caused by the gradual removal of top soil by natural processes which bring equilibrium between physical, biological and hydrological activities and maintain a natural balance between erosion and renewal.

Accelerate erosion: This is mainly caused by man-made activities and the rate of erosion is much faster than the rate of formation of soil. Overgrazing, deforestation and mining are some important activities causing accelerated erosion

There are two types of agents which cause soil erosion. They are climatic agents and biotic agents.

Climatic Agents – Water and Wind:

Water affects soil erosion in the form of rain. **Water-induced soil erosion** is of following types:

Sheet erosion: When there is the uniform removal of a thin layer of soil from a large surface area, it is called sheet erosion.

Rill erosion: when there is rainfall and rapidly running water produces finger-shaped grooves or rolls over the area, it is called rill erosion.

Gully erosion: When the rainfall is very heavy, deeper cavities or gullies are formed, which may be U or V shaped.

Slip erosion: This occurs due to heavy rainfall on slopes of hills and mountains.

Stream bank erosion: During the rainy season, when fast running streams take a turn in some other direction, they cut the soil and make caves in the bank

Wind erosion is responsible for the following three types of soil movements:

Saltation: This occurs under the influence of the direct pressure of stormy wind and the soil particles of 1-1.5 mm diameter move up in vertical direction.

Suspension: Here fine soil particles (less than 1mm diameter) which are suspended in the air are picked up and taken away to distant places.

Surface creep: Here the large particles (5-10 mm diameter) creep over the soil surface along with the wind.

Biotic Agents

- Excessive grazing, mining, and deforestation are the major biotic agents responsible for soil erosion.
- Deforestation without deforestation, overgrazing by cattle, surface mining without land reclamation, irrigation techniques that lead to salt build- up, waterlogged soil, make the topsoil vulnerable to erosion.

Soil Conservation Practices

In order to prevent soil erosion and conserve the soil, the following practices are employed.

Conservational Till Farming

In the traditional method, the soil is broken up and smoothed to make a planting surface. This disturbs the soil and makes it susceptible to erosion.

Conservational till farming, popularly known as no-till-farming cause's minimum disturbance to the top soil.

Here special tillers break up and loosen the subsurface soil without turning over the topsoil.

The tilting machines make slits in the soil and inject seeds, fertilizers, and a little water in the slit so that crop grows successfully.

Contour Farming

On gentle slopes, crops are grown in rows across, rather up and down. This practice is known as contour farming. It helps to hold soil and slow down the loss of soil through run-off water.

Terracing

It is used on still steeper slopes are converted into a series of broad terraces which run across the contour. Terracing retains water for crops at all levels and cuts down soil erosion.

Strip Cropping

Here strips of crops are alternated with strips of soil saving crops like grasses or grass-legume mixture. Whatever run-off comes from the cropped soil is retained by the strip of cover- crop and this reduces soil erosion.

Alley Cropping

It is a form of inter – cropping in which crops are planted between rows of trees or shrubs. This is also called **Agroforestry**. Even when the crop is harvested, the soil is not fallow because trees and shrubs still remain on the soil holding the soil particles and prevent soil erosion.

Plate I(a) Terrace farming

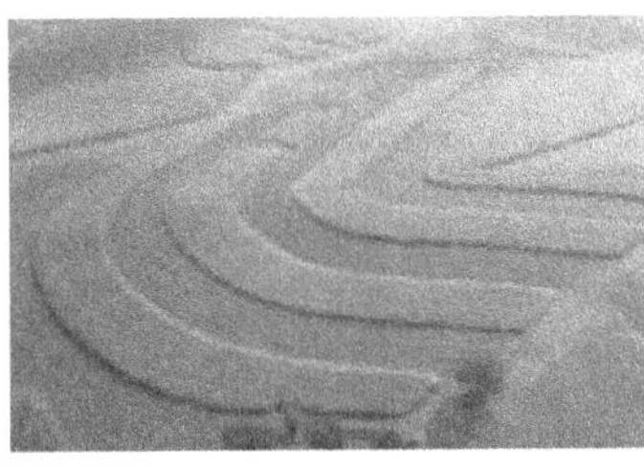

Plate I(b) Strip cropping

Plate I(c) Alley cropping

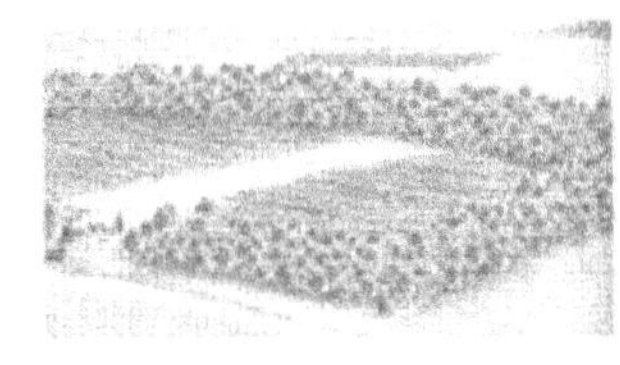

Plate I(d) Shelter belt

Windbreaks or Shelterbelts

The trees are planted in long rows along the cultivated land boundary so that the wind is blocked. The wind speed is substantially reduced which helps in preventing wind erosion of soil.

Desertification

Desertification is characterized by vegetation and loss of vegetal cover, depletion of groundwater, salinization, and severe soil erosion.

Desertification leads to the conversion of irrigated croplands to desert-like conditions in which agricultural productivity falls. Moderate desertification produces 10-25% drop in productivity. Severe desertification causes 25-50% drop while very severe desertification results in 50% drop in productivity.

Causes of Desertification: The major man-made activities responsible for desertification are as follows.

Deforestation

- Deforestation means the destruction of forests.
- The total forests area of the world in 1900 was estimated to be 7,000 million hectares which were reduced to 2890 million ha in 1975 fell down to just 2,300 million ha by 2000.
- Deforestation rate is relatively less in temperature countries, but it is very alarming in tropical countries.

Overgrazing

- Overgrazing can limit livestock production.
- Overgrazing occurs when too many animals graze for too long and exceed the carrying capacity of a grassland area.
- Overgrazing removes the grass cover.
- The humus content of the soil is decreased and it leads to poor, dry, compacted soil.
- The soil roots are very good binders of soil.
- When the grasses are removed, the soil becomes loose and susceptible to the action of wind and water.
- The dry barren land reflects more of the sun's heat, changing wind patterns leading to further desertification.

Mining and Quarrying

Mining operation requires removal of vegetation along with underlying soil mantle and overlying rock masses.

This results in the destruction of the landscape in the area.

1.8.   Role of Individual in Conservation of Natural Resources

Different natural resources like forests, water, soil, food, mineral and energy resources play a vital role in the development of a nation. While conservation efforts are underway at National as well as International level, the individual efforts for conservation of natural resources can go a long way.

Conserve Water

- Don't keep water taps running while brushing, shaving, washing or bathing.
- Check for water leaks in pipes and toilets and repair them promptly. A small pin-hole sized leak will lead to the wastage of 640 liters of water in a month.
- Use drip irrigation and sprinkling irrigation to improve irrigation efficiency and reduce evaporation.
- Install a small system to capture rainwater and collect normally wasted used water from sinks, cloth-washers, bathtubs etc. which can be used for watering the plants
- Build rainwater harvesting system in your house. Even the President of India is doing this.

Conserve Energy

- Turn off lights, fans and other appliances when not in use.
- Obtain as much heat as possible from natural sources. Dry the clothes in the sun instead of drier if it is a sunny day.
- Use solar cooker for cooking your food on sunny days which will be more nutritious and will cut down on your LPG expenses.
- Grow deciduous trees and climbers at proper places outside your home to cut off the intense heat of summers and get a cool breeze and shade. This will cut off your electricity charges on coolers and air-conditioners.
- Try riding a bicycle or just walk down small distances instead of using your car or scooter.

Protect the Soil

- While constructing your house, don't uproot the trees as far as possible. Plant the disturbed areas with a fast growing native ground cover.
- Make compost from your kitchen waste and use it for your kitchen-garden or flower-pots.
- Do not irrigate the plants using a strong flow of water, as it would wash off the soil.

- If you own agricultural fields, do not over-irrigate your fields without proper drainage to prevent water logging and salinization.
- Use mixed cropping so that some specific soil nutrients do not get depleted.

Promote Sustainable Agriculture

- Do not waste food. Take as much as you can eat.
- Reduce the use of pesticides.
- Fertilize your crop primarily with organic fertilizers.
- Eat local and seasonal vegetables. This saves a lot of energy on transport, storage and preservation.
- Control pests by a combination of cultivation and biological control methods.

Equitable Use of Resources for Sustainable Life Style

- There is a big divide in the world as North and South, the more developed countries (MDC'S) and less developed countries (LDC'S), the haves and the have-nots.
- The MDC's have only 22% of world's population, but they use 88% of its natural resources, 73% of its energy and command 85% of its income.
- As the rich nations continue to grow, they will reach a limit.
- If they have a growth rate of 10% every year, they will show 1024 times increase in the next 70 years.
- Will this much of growth be sustainable? The answer is 'No' because many of our earth's resources are limited and even the renewable resources will become unsustainable if their use exceeds their regeneration.
- Thus, the solution to this problem is to have more equitable distribution of resources and wealth.
- We cannot expect the poor countries to stop growth in order to check pollution because development brings employment and the main problem of these countries is to tackle poverty.
- The poor in the LDC'S are at least able to sustain their life.
- Unless they are provided with such basic resources, we cannot think of rooting out the problems related to dirty, unhygienic, polluted, disease infested settlements of these people which contribute to unsustainability.
- Thus, the two basic causes of unsustainability are an overpopulation in poor countries who have under consumption of resources and over-consumption of resources by the rich countries, which generate wastes.

- In order to achieve sustainable lifestyles, it is desirable to achieve a more balanced and equitable distribution of global resources and income to meet everyone's basic needs.
- The rich countries will have to lower down their consumption levels while the bare minimum needs of the poor have to be fulfilled by providing them resources.
- A fairer sharing of resources will narrow down the gap between the rich and the poor and will lead to sustainable development for all and not just for a privileged group.

A field study of the local area to document environmental assets–river/forest/grassland/hill/mountain.

Unit II

Ecosystem and Biodiversity

Various kinds of life supporting systems like the forests, grasslands, oceans, lakes, rivers, mountains show wide variations in their structural composition and functions. However, it is a fact that it interacts with their surroundings exchanging matter and the energy

2.1. Concept of an Ecosystem

- The term Ecology was coined by Earnst Haeckel in 1869. It is derived from the Greek words Oikos- home + logos- study.
- **Ecology** deals with the study of organisms in their natural home interacting with their surroundings.
- An ecosystem is a group of biotic communities of species interacting with one another and with their non-living environment exchanging energy and matter.
- An **ecosystem** is a unit or a system which is composed of a number of subunits that are all directly or indirectly linked with each other.
- They may be freely exchanging energy and matter from outside—an open ecosystem or may be isolated from outside—a closed ecosystem.

Structure of an Ecosystem

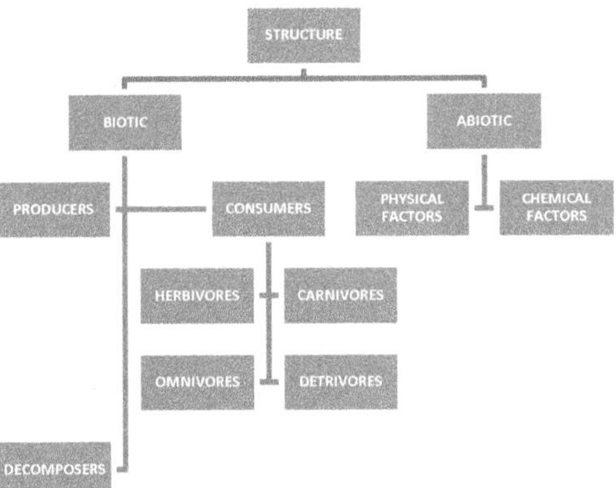

Ecosystems show large variations in their size, structure, composition. However, all the ecosystems are characterized by certain basic structural and functional features which are common as listed. The structure of an ecosystem explains the relationship between the abiotic (nonliving) and the biotic (living) components.

Biotic Structure

The plants, animals and microorganisms present in an ecosystem from the biotic component.

These organisms have different nutritional behaviour and status in the ecosystems and are accordingly known as Producers or Consumers, based on how they get their food.

Producers

- Producers are mainly the green plants, which can synthesize their food themselves by making use of carbon dioxide present in the air and water in the presence of sunlight by involving chlorophyll, the green pigment present in the leaves, through the process of photosynthesis.
- They are also known as **photo autotrophs** (auto=self; troph=food, photo=light).
- There are some microorganisms also which can produce organic matter to some extent through oxidation of certain chemicals in the absence of sunlight. They are known as **chemosynthetic organisms** or chemoautotrophs.
- For instance in the ocean depths, where there is no sunlight, chemoautotrophic sulphur bacteria make use of the heat generated by the decay of radioactive elements present in the earth's core and released in ocean's depths.
- They use this heat to convert dissolved hydrogen sulphide (H_2S) and carbon dioxide (CO_2) into organic compounds.

Consumers

All organisms which get their organic food by feeding upon other organisms are called consumers, which are of the following types.

1. **Herbivores (plant eaters):** They feed directly on producers and hence also known as primary consumers. e.g. rabbit, insect, man.
2. **Carnivores (meat eaters):** They feed other consumers. If they feed on herbivores they are called secondary consumers (e.g. frog) and if they feed on the carnivores (snake, big fish etc.) they are known as tertiary consumers.
3. **Omnivores:** They feed on both plants and animals. E.g. humans, rat, fox, many birds.
4. **Detritivores (Detritus feeders or Saprotrophs):** They feed on the parts of dead organisms, wastes of living organisms, their castoffs and partially decomposed matter e.g. beetles, termites, ants, crabs, earthworms etc.

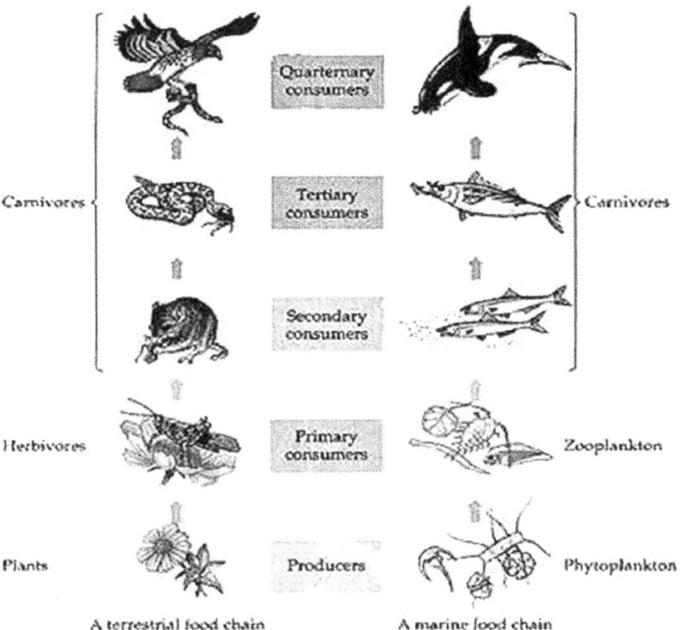

Decomposers

Decomposers derive their nutrition by breaking down the complex organic molecules to simpler organic compounds and ultimately into inorganic nutrients. Ex: Various bacteria and fungi are decomposers.

In all the ecosystems, this biotic structure prevails. However, in some, it is the primary producers which predominate (e.g. in forests, agroecosystems) while in others the decomposers predominate (e.g. deep ocean).

Abiotic Structure

The physical and chemical components of an ecosystem constitute its abiotic structure. It includes climatic factors, edaphic (soil) factors, geographical factors, energy, nutrients and toxic substances.

Physical Factors

Average temperature, sunlight, duration of sunlight, the wind, latitude all comes under physical factors.

Chemical Factors

Major essential nutrients such as nitrogen, phosphorous, hydrogen, oxygen, level of toxic substances.

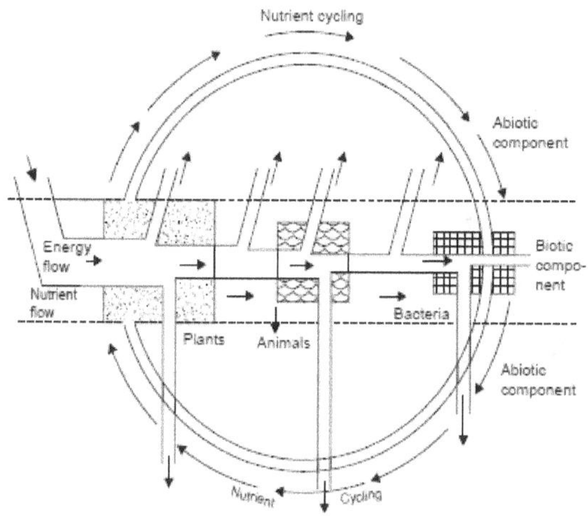

Functional Attributes

Every ecosystem performs under natural conditions in a systematic way. It receives energy from the sun and passes it on through various biotic components and in facts, all life depends on upon this flow of energy. The major functional attributes of an ecosystem are as follows:

1) Food chain, food webs and tropic structure.
2) Energy flow.
3) Cycling of nutrients (Biogeochemical cycles).
4) Primary and Secondary production.
5) Ecosystem development and regulation.

Food Chain

The sequence of eating and being eaten in an ecosystem is known as **food chain.**

- All organisms, living or dead, are potential food for some other organism and thus, there is essentially no waste in the functioning of a natural ecosystem.

- A caterpillar eats a plant leaf, a sparrow eats the caterpillar, a cat or a hawk eats the sparrow and when they all die, they are all consumed by microorganism like bacteria or fungi (decomposers) which break down the organic matter and convert it into simple inorganic substances that can again be used by the plants the primary producers.

Some common examples of simple food chains are:

- **Grass → grasshopper→ Frog → Snake →Hawk (Grassland ecosystem)**
- **Phytoplanktons → water fleas → small fish → Tuna (Pond ecosystem)**
- **Lichens → reindeer → Man (Arctic tundra)**

- Each organism in the ecosystem is assigned a feeding level or trophic level depending on its nutritional status.
- Thus, in the grassland food chain, grasshopper occupies the I tropic level, frog the II and snake and hawk occupy the III and the IV tropic levels, respectively.

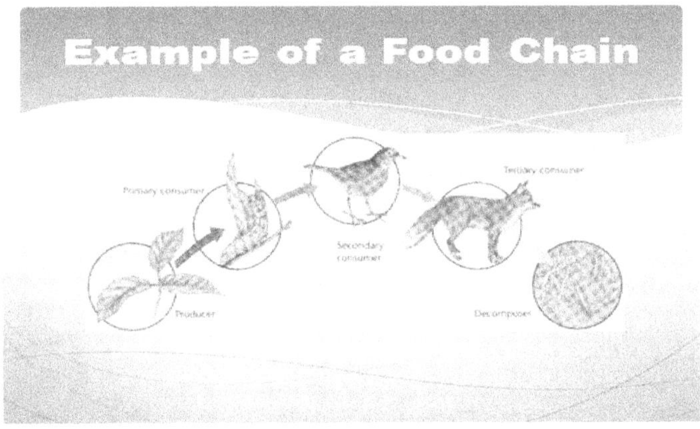

In nature, we come across two major types of food chains:

1. **Grazing food chain:** It starts with green plants (primary producers) and culminates in carnivores. Example: Grass→ Rabbit→ Fox.

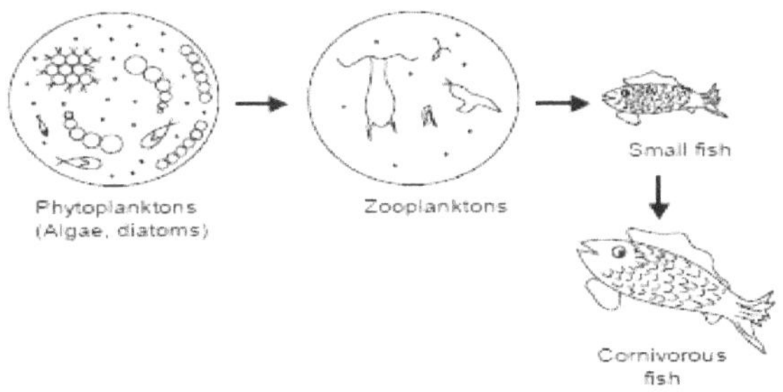

A Grazing Food chain in a Pond Ecosystem

2. **Detritus food chain:** It starts with a dead organic matter which the detritivores and decomposers consume. Partially decomposed dead organic matter and even the decomposers are consumed by detritivores and their predators.

Examples: Leaf litter→ algae→ crabs→ small carnivorous fish→ large carnivorous fish (Mangrove ecosystem) Dead organic matter→ fungi→ bacteria (Forest ecosystem).

Both the food chains occur together in natural ecosystems, but grazing food chain usually predominates.

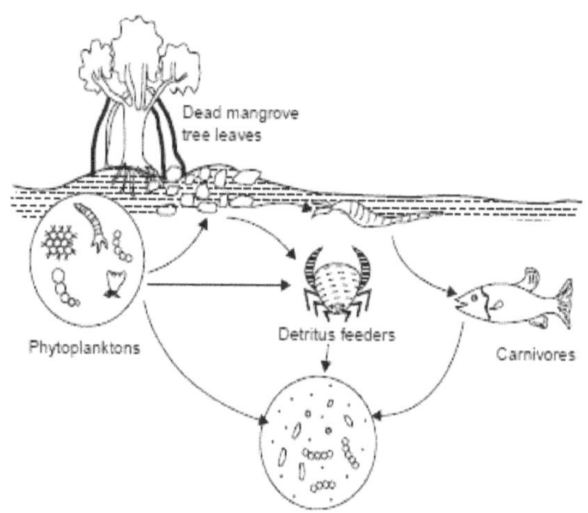

Food Web

A food web is a network of food chains where different types of organisms are connected at the different tropic level so that there are a number of options of eating and being eaten at each tropic level.

- In a tropical region, the ecosystems are much more complex.
- They have rich species diversity and therefore, the food webs are much more complex.
- Food webs give greater stability to the ecosystem.
- In a linear food chain, if one species becomes extinct or one species suffers then the species in the subsequent trophic levels are also affected.
- In a food web, on the other hand, there are a number of options available at each trophic level.
- So if one species is affected, it does not affect other trophic levels so seriously.

For Example, Hawk eats both mice and birds. Coyote eats mice, rabbits and birds.

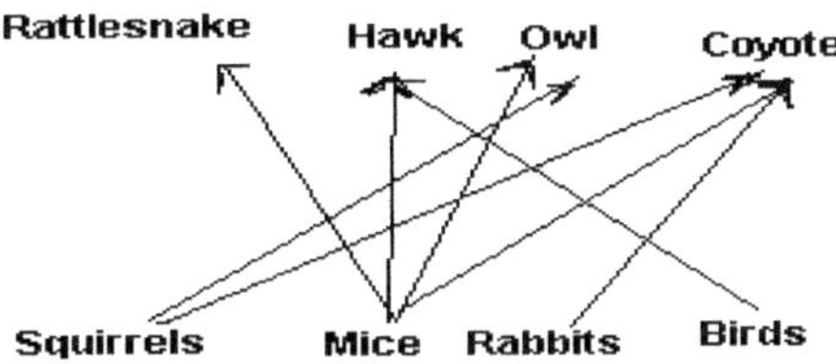

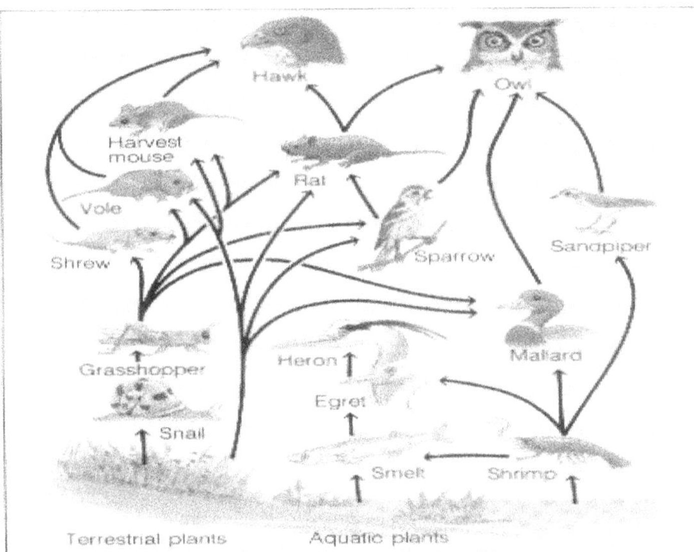

Significance of Food Chains and Food Webs

- Food chains and food webs play a very significant role in the ecosystem because the two most important functions of energy flow and nutrient cycling take place through them.
- They help maintain the ecological balance.
- Food chains show a unique property of biological magnification of some chemicals.

Energy Flow in an Ecosystem

The flow of energy in an ecosystem takes place through the food chain and it is this energy flow which keeps the ecosystem going. The most important feature of this energy flow is that it is unidirectional or one-way flow. Unlike the nutrients, (like carbon, nitrogen, phosphorus etc.) energy is not reused in the food chain.

Also, the flow of energy follows the two laws of Thermodynamics:

I law of thermodynamics states that energy can neither be created nor be destroyed but it can be transferred from one form to another. The solar energy captured by the green plants (producers) gets converted into biochemical energy of plants and later into that of consumers.

II law of Thermodynamics states that energy dissipates as it is used or in other words, it gets converted from a more concentrated to dispersed form. As energy flows through the food chain, there occurs dissipation of energy at every trophic level.

2.2. Nutrient Flow (OR) Biogeochemical Cycles

Cyclic flow of nutrients between the biotic and abiotic components is known as nutrient cycle.

Carbon Cycle

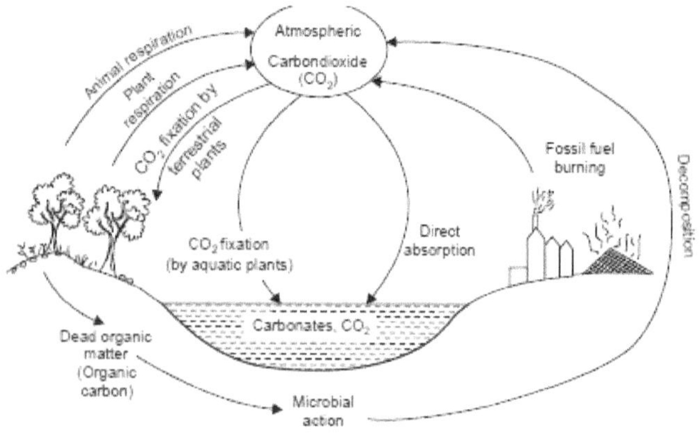

Nitrogen Cycle

Nitrogen is essential to life because it is a key component of proteins and nucleic acids. Nitrogen occurs in many forms and is continuously cycled among these forms by a variety of bacteria. Although nitrogen is abundant in the atmosphere as diatomic nitrogen gas (N2), it is extremely stable, and conversion to other forms requires a great deal of energy. Historically, the biologically available forms NO3- and NH3 have often been limited; however, current anthropogenic processes, such as fertilizer production, have greatly increased the availability of nitrogen to living organisms. The cycling of nitrogen among its many forms is a complex process that involves numerous types of bacteria and environmental conditions.

In General, the Nitrogen Cycle has Five Steps

1. Nitrogen fixation (N_2 to NH_3/ NH_4+ or NO^{3-}).
2. Nitrification (NH_3 to NO^{3-}).
3. Assimilation (Incorporation of NH_3 and NO^{3-} into biological tissues).
4. Ammonification (organic nitrogen compounds to NH_3).
5. Denitrification (NO^{3-} to N_2).

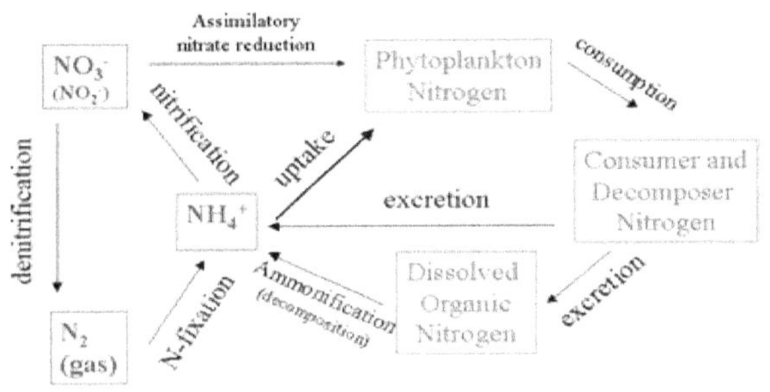

Nitrogen Fixation

Nitrogen fixation is the process by which gaseous nitrogen (N2) is converted to ammonia (NH3 or NH4+) via biological fixation or nitrate (NO3-) through high-energy physical processes. N2 is extremely stable and a great deal of energy is required to break the bonds that join the two N atoms.

N2 can be converted directly into NO3- through processes that exert a tremendous amount of heat, pressure, and energy. Such processes include combustion, volcanic action, lightning discharges, and industrial means. However, a greater amount of biologically available nitrogen is naturally generated via the biological conversion of N2 to NH3/ NH4+. A small group of bacteria and cyanobacteria are capable using the enzyme nitrogenase to break the bonds among the molecular nitrogen and combine it with hydrogen.

Nitrification

Nitrification is a two-step process in which NH3/ NH4+ is converted to NO3-. First, the soil bacteria Nitrosomonas and Nitrococcus convert NH3 to NO2-, and then another soil bacterium, Nitrobacter, oxidizes NO2- to NO3-. These bacteria gain energy through these conversions, both of which require oxygen to occur.

Assimilation

Assimilation is the process by which plants and animals incorporate the NO3- and ammonia formed through nitrogen fixation and nitrification.

Plants take up these forms of nitrogen through their roots and incorporate them into plant proteins and nucleic acids. Animals are then able to utilize nitrogen from the plant tissues.

Ammonification

Assimilation produces large quantities of organic nitrogen, including proteins, amino acids, and nucleic acids.

Ammonification is the conversion of organic nitrogen into ammonia. The ammonia produced by this process is excreted into the environment and is then available for either nitrification or assimilation.

Denitrification

Denitrification is the reduction of NO3- to gaseous N2 by anaerobic bacteria. This process only occurs where there is little to no oxygen, such as deep in the soil near the water table. Hence, areas such as wetlands provide a valuable place for reducing excess nitrogen levels via denitrification processes.

2.3.   Ecological Pyramids

Introduction

Graphic representation of trophic structure and function of an ecosystem, starting with producers at the base and successive trophic levels forming the apex is known as an **ecological pyramid**.

Types

Ecological pyramids are of three types:

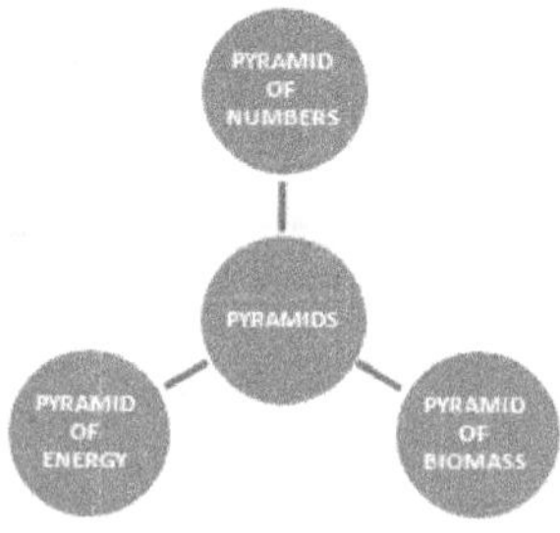

2.4.   Characteristic Features

1. *Pyramid of Numbers*

It represents the number of individual organisms at each trophic level. We may have upright or inverted pyramid of numbers, depending upon the type of ecosystem and food chain as shown in the figure. A grassland ecosystem and a pond ecosystem show an upright pyramid of numbers. The producers in the grasslands are grasses and that in a pond are phytoplankton (algae etc.), which are small in size and very large in number. So the producers from a broad base. The herbivores in grassland are insects while tertiary carnivores are hawks or other birds which are gradually less and less in number and hence the pyramid apex becomes gradually narrower forming an upright pyramid.

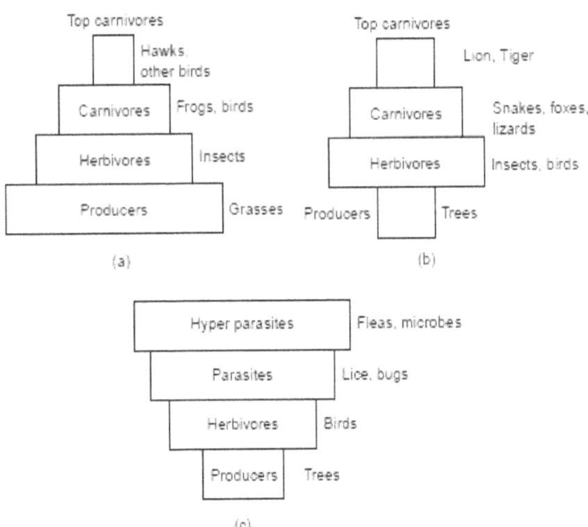

Pyramid of numbers (a) Grassland (b) Forest (c) Parasitic Food Chain

2. *Pyramid of Biomass*

It is based on the total biomass (dry matter) at each trophic level in a food chain. The pyramid of biomass can also be upright or inverted as in figure which shows pyramids of biomass in an aquatic ecosystem.

The pond ecosystem shows an inverted pyramid of biomass .The total biomass of producers (phytoplanktons) is much less as compared to herbivores (zooplanktons, insects), carnivores (Small fish) and tertiary carnivores (big fish). Thus, the pyramid takes an inverted shape with a narrow base and broad apex.

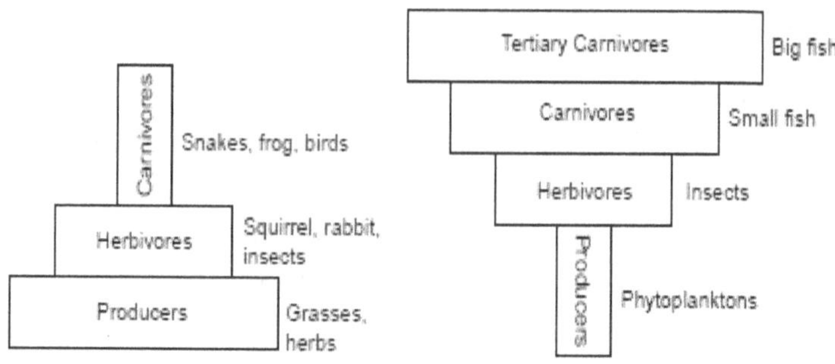

Pyramid of Biomass (a) Grassland (b) Pond

3. Pyramid of Energy

The amount of energy present at each trophic level is considered for this type of pyramid. It gives the best representation of the tropic relationships and it is always upright. There is a sharp decline in the energy level of each successive trophic level as we move from producers to top carnivores. Therefore, the pyramid of energy is always upright as shown in the figure.

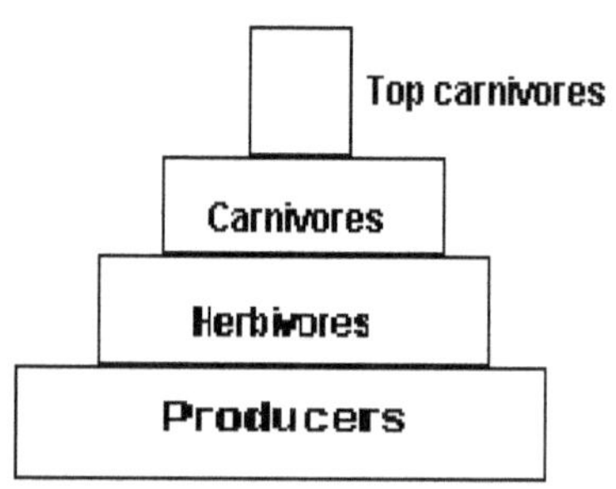

Pyramid of Energy

2.5. Major Types of Ecosystem

Forest Ecosystem

Depending on the climate conditions, forest may be classified as:

a) **Tropical Rain Forests**: They are evergreen broadleaf forests found near the equator. They are characterized by high temperature, high humidity, and high rainfall, all of which favor the growth of trees. Different layers of plants and animals are also dealt under tropical rain forests. The emergent layer is the top most layer of the broad leaves below which comes the canopy where the top branches look like an umbrella below which comes understory layer. On the tree trunks, some woody climbers are found to grow known as lianas. There are some other plants like orchids which are epiphytes.

The understory trees usually receive less amount of sunlight. They usually develop dark green leaves with high chlorophyll content so that they use the diffused sunlight for photosynthesis. The shrub layer receives even less sunlight and the ground layer commonly known as forest floor receives almost no sunlight and it is a dark layer. If nutrients present it will be accumulated by the my corrhizal roots of the trees.

b) **Tropical deciduous forests**: They are found a little away from the equator and are characterized by a warm climate the year round. Rain occurs only during monsoon. Different types of deciduous trees are found here which lose their leaves during the dry season.

c) **Tropical scrub forests**: They are found in areas where the day season is even longer.

d) **Temperate rain forests**: They are found in temperate areas with adequate rainfall. These are dominated by trees like pines, firs, redwoods etc.

e) **Temperate deciduous forests**: They are found in areas with moderate temperatures. It is characterized by long summers, cold but not severe but abundant rainfall throughout the year.

f) **Evergreen coniferous forests (Boreal Forests)**: They are found just south of arctic tundra. Here winters are long, cold and dry. Sunlight is available for a few hours only. The major trees include pines, spruce, fir, cedar. The soil found in this region usually gets frozen during winter.

The abiotic environment of forest ecosystem includes the nutrients present in the soil in forest floor which is usually rich in dead and decaying organic matter.

Producers: Producers are mainly big trees, some shrubs and ground vegetation.

Primary consumers: Primary consumers are insects like ants, flies, beetles, spiders, and big animals like elephants, deer, squirrels etc.

Secondary consumers: Secondary consumers are carnivores like snakes, lizards, foxes, birds etc.,

Tertiary consumers: Tertiary consumers are animals like tiger, lion etc.

Decomposers: Decomposers are bacteria fungi which are found in soil on the forest floor. The rate of decomposition in tropical or sub-tropical forests is more rapid than that in the temperate zones.

Grassland Ecosystem

The grassland ecosystem occupies about 10% of the earth's surface. The abiotic environment includes nutrient like nitrates, sulphates or phosphates and trace elements present in the soil, gases, like CO_2 present in the atmosphere and water etc. Limited grazing helps to improve the net primary production of the grasslands but overgrazing leads to the destruction.

Three types of grasslands are found to occur in different climatic regions:

(a) **Tropical grasslands**: They occur near the borders of tropical rain forests in regions of high average temperature and low to moderate rainfall. In Africa, these are typically known as savannas which have a wide diversity of animals such as zebras, giraffe, and gazelle. During dry season fire is quite common. Tropical savannas have a highly efficient system of photosynthesis. Most of the carbon assimilated by them is in the form of carbohydrates which are pertaining in the underground.

(b) **Temperate grasslands**: They are usually found on flat, gentle sloped hills, winters are very cold but summers are hot and dry. In the United States and Canada, these grasslands are known as prairies, in South America as pampas, in Africa as Velds, in Asia known steppes.

(c) **Polar grasslands**: they are found in the arctic polar region where severe cold and strong, frigid winds along with ice and snow create too harsh a climate for trees to grow. A thick layer of ice remains under the soil in frozen form throughout the year known as permafrost.

Producers: Producers are mainly grass and some herbs, shrubs, and few scattered trees.

Primary consumers: Primary consumers are grazing animals such as cow, sheep, deer, house, kangaroo, etc. Some insects and spiders have also been included as primary consumers.

Secondary consumers: Secondary consumers are animals like fox, jackals, snakes, lizards, frogs and birds etc.

Tertiary consumers: Decomposers are bacteria, moulds, and fungi, like penicillium, Aspergillums etc. The minerals and other nutrients are thus brought back to the soil and are made available to the producers.

Desert Ecosystem

Desert occurs in the region where the average rainfall is less than 25 cm. The abiotic environment of a desert ecosystem includes water which is scarce. The atmosphere is very very dry and hence, it is a poor insulator. That is why in deserts the soil gets cooled up quickly, making the nights cool.

Deserts are three major types, based on climatic conditions:

1. **Tropical deserts** like Sahara in Africa and the Thar Desert, Rajasthan, India are the driest of all with only a few species.
2. **Temperature deserts** like Mojave in Southern California where daytime temperatures are very hot in summer but cool in winters.
3. **Cold deserts** like Gobi desert in China have cold winters and warm summers.

Producers: the chief producers are shrubs, bushes, and some trees whose roots are very extensive and stems and leaves are modified to store water and to reduce the loss of water as a result of transpiration. Low plants such as mosses and blue-green algae are minor producers.

Primary consumers: Primary consumers are animals like rabbits which get water from succulent plants. They do not drink water even if it is freely available. Camel is also a primary consumer of the desert.

Secondary consumers: Secondary consumers are carnivores like reptiles having impervious skin which minimize loss water from the surface of the body.

Tertiary consumers: The tertiary consumers are mainly birds which conserve water by excreting solid uric acid.

Decomposers: Decomposers are bacteria and fungi which can thrive in hot climate conditions. Because of the scarcity of flora and fauna, the dead organic matter available is much less and therefore, decomposers are also less in number.

Aquatic Ecosystems

Aquatic ecosystems dealing with water bodies and the biotic communities present in them are either freshwater or marine. Freshwater ecosystems are of standing type (Lentic) or free flowing like (Lottic). Let us consider some important aquatic ecosystems.

1. *Pond Ecosystems*

- It is a small freshwater aquatic ecosystem where water is stagnant.
- Ponds may be seasonal in nature i.e. receiving enough water during rainy season.
- Ponds are usually shallow water bodies which play a very important role in the villages where most of the activities center around ponds.
- They contain several types of algae, aquatic plants, insects, fishes, and birds.
- The ponds are, however, very often exposed to tremendous anthropogenic pressures.
- They are used for washing clothes, bathing, swimming, cattle bathing and drinking etc. and therefore get polluted.

2. *Lake Ecosystems*

- Lakes are usually big freshwater bodies with standing water.
- They have shallow water zone called **Littoral zone**, an open-water zone called **Limnetic zone** and deep bottom area where light penetration is negligible, known as pro fundal zone.

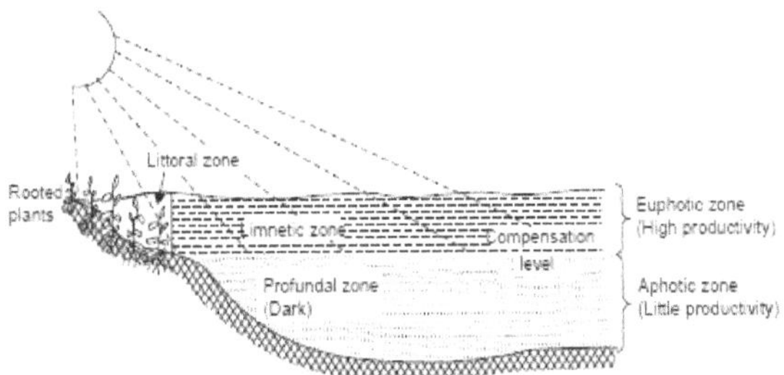

Zonation in a Lake Ecosystem

Organisms: Lakes have several types of organisms:

- **Planktons** that float on the surface of waters e.g. phytoplanktons like algae and zooplanktons like rotifers.
- **Nektons** that swim e.g. fishes.
- **Neustons** that rest or swim on the surface.

- **Benthos** that are attached to bottom sediments e.g. snails.
- **Periphytons** that are attached or clinging to other plants or any other surface e.g. crustaceans.

Stratification

The lakes show stratification or zonation based on temperature differences. During summer, the top waters become warmer than the bottom waters. Therefore, only the warm top layer circulates without mixing with the colder layer, thus forming a distinct zonation

Epyilimnion: Warm, lighter, circulating surface layer.

Hypolimnion: Cold, viscous, non-circulating bottom layer

In between the above two layers is thermocline, a region of a sharp drop in temperature.

Types of Lakes

- **Oligotrophic lakes** which have low nutrient concentrations.
- **Eutrophic lakes** which are over nourished by nutrients like nitrogen and phosphorus, usually as a result of agricultural run-off or municipal sewage discharge. They are covered with "algal blooms" e.g. Dal lake.
- **Dystrophic lakes** that have low pH, high humic acid content and brown waters e.g. bog lakes.
- **Endemic lakes** that are very ancient, deep and have endemic fauna which is restricted only to that lake e.g. the Lake Baikal in Russia.
- **Artificial lakes or impoundments** that are created due to the construction of dams e.g. Govind Sagar Lake at Bhakra-Nangal.

3. Streams

These are freshwater aquatic ecosystems where water current is a major controlling factor, oxygen and nutrient in the water are more uniform and land-water exchange is more extensive. Although stream organisms have to face more extremes of temperature and action of currents as compared to pond or lake organisms, but they do not have to face oxygen deficiency under natural conditions. This is because the streams are shallow, have a large surface exposed to air and constant motion which churns the water and provides abundant oxygen. Their dissolved oxygen level is higher than that of ponds even though the green plants are much less in number. The stream animals usually have a narrow range of tolerance to oxygen. That is the reason why they are very susceptible to any organic pollution which depletes dissolved oxygen in the water. Thus, streams are the worst victims of industrial development.

4. River Ecosystems

Rivers are large streams that flow downward from mountain highlands and flowing through the plains fall into the sea. So the river ecosystems show a series of different conditions.

The mountain highland part has cold, clear waters rushing down as waterfalls with large amounts of dissolved oxygen.

In the second phase on the gentle slopes, the waters are warmer and support a luxuriant growth of plants and less oxygen requiring fishes.

In the third phase, the river waters are very rich in biotic diversity. Moving down the hills, rivers shape the land. They bring with them lots of silt rich in nutrients which are deposited in the plains and in the delta before teaching the ocean.

5. Oceans

These are gigantic reservoirs of water covering more than 70% of our earth's surface and play a key role in the survival of about 2,50,000 marine species, serving as food for humans and other organisms, give a huge variety of sea-products and drugs. Oceans provide us iron, phosphorus, magnesium; oil, natural gas, sand, and gravel. Oceans are the major sinks of carbon dioxide and play an important role in regulating many biogeochemical cycles and hydrological cycle, thereby regulating the earth's climate. The oceans have two major life zones

Coastal zone: It is relatively warm, nutrient rich shallow water. Due to high nutrients and ample sunlight, this is the zone of high primary productivity.

Open sea: It is the deeper part of the ocean, away from the continental shelf. It is vertically divided into three regions:

- **The euphotic zone** which receives abundant light and shows high photosynthetic activity.
- **Bathyal zone** receives dim light and is usually geologically active.
- **Abyssal zone** is the dark zone, 2000 to 5000 meters deep. The abyssal zone has no primary source of energy i.e. solar energy. It is the world's largest ecological unit but it is an incomplete ecosystem.

6. Estuary

Estuary is a partially enclosed coastal area at the mouth of a river where fresh water and salty seawater meet. These are the transition zones which are strongly affected by the tidal action. Constant mixing of water stirs up the silt which makes the nutrients available for the primary producers.

The organisms present in estuaries show a wide range of tolerance to temperature and salinity. Such organisms are known as eurythermal and euryhaline. Coastal bays and tidal marshes are examples of estuaries. Estuary has a rich biodiversity and many of the species are endemic.

There are many migratory species of fishes like eels and salmons in which half of the life is spent in the fresh water and a half in the salty water. For them, estuaries are ideal places for resting during migration, where they also get abundant food. Estuaries are highly productive ecosystems. The river flow and tidal action provide energy for estuary thereby enhancing its productivity.

Estuaries are of much use to human beings due to their high food potential. However, these ecosystems need to be managed judiciously and protected from pollution.

2.6. Bio Diversity

World Earth Day-April 22nd

Introduction to Biodiversity

Definition

Biodiversity refers to the variety and variability among all groups of living organisms and the ecosystem complexes in which they occur.

In the convention of Biological diversity (1992) biodiversity has been defined as the variability among living organisms from all sources including inter alia, terrestrial, marine and other aquatic ecosystems and the ecological complexes of which they are a part.

Levels of Biodiversity

Genetic Diversity

Genetic Diversity is the basic source of biodiversity. The genes found in organisms can form an enormous number of combinations each of which gives rise to some variability. Genes are the basic units of hereditary information transmitted from one generation to other. When the genes within the same species show different versions due to new combinations, it is called genetic variability.

For example, all rice varieties belong to the species Oryza Sativa, but there are thousands of wild and cultivated verities of rice which show variations at the genetic level and differ in their color, size, shape, aroma and nutrient content of the grain. This is the genetic diversity of rice

Species Diversity

Species Diversity is the variability found within the population of a species or between different species of a community. It represents broadly the species richness and their abundance in a community. There are two popular indices of measuring species diversity known as *Shannon-wiener index* and *Simpson index.*

What is the number of species in this biosphere?

- The estimates of actual number vary widely due to incomplete and indirect data.
- The current estimates given by Wilson in 1992 put the total number of living species in a range of 10 million to 50 million.
- Till now only about 1.5 million living and 300,000 fossil species have been actually described and given scientific names.

Ecosystem Diversity

Ecosystem diversity is the diversity of ecological complexity showing variations in trophic structure, food webs, nutrient cycling etc. The ecosystems also show variations with respect to physical parameters like moisture, temperature, altitude, precipitation etc.

The ecosystem diversity is of great value that must be kept intact. This diversity has developed over millions of years of evolution. If we destroy this diversity, it would disrupt the ecological balance. We cannot even replace the diversity of one ecosystem by that of another. Coniferous trees of boreal forests cannot take up the function of the trees of tropical deciduous forest lands and vice versa.

Value of Biodiversity

The value of biodiversity in terms of its commercial utility, ecological services, the social and aesthetic value is enormous.

The multiple uses of biodiversity value have been classified by McNeely et al in 1990 as follows:

(i) Consumptive use value: these are direct use values where the biodiversity product can be harvested and consumed directly e.g. fuel, food, drugs, fibre etc.

    a) Food: A large number of wild plants are consumed by human beings as food. About 80,000 edible plant species have been reported from wild. About 90% of present day food crops have been domesticated from wild tropical plants. A large number of wild animals are also our sources of food.

b) Drugs and medicines:

1. About 75% of the world's population depends on upon plants or plant extracts for medicines.
2. The wonder drug penicillin used as an antibiotic is derived from a fungus called penicillium.
3. Likewise, we get Tetracyclin from a bacterium. Quinine, the cure for malaria is obtained from the bark of Cinchona tree while Digitalin is obtained from foxglove which is an effective cure for heart ailments.
4. Recently vinblastine and vincristine, two anticancer drugs, have been obtained from a periwinkle plant, which possesses anticancer alkaloids.

Our forests have been used since ages for fuel wood. The fossil fuels coal, petroleum, and natural gas are also products of fossilized biodiversity.

(ii) Productive use Values

a) These are the commercially usable values where the product is marketed and sold.
b) These may include the animal products like tusks of elephants, musk from musk deer, silk from the silkworm, wool from sheep, etc, all of which are traded on the market.
c) Many industries are dependent upon the productive use values of biodiversity e.g. – the paper and pulp industry, plywood industry, railway sleeper industry, silk industry, ivory-works, leather industry, pearl industry etc.

(iii) Social Value

a) These are the values associated with the social life, customs, and religion of the people.
b) Many of the plants are considered holy and sacred in our country like Tulsi, peepul, Mango, and Lotus etc.
c) The leaves, fruits or flowers of these plants are used in worship or the plant itself is worshipped.
d) Many animals like Cow, Snake, and Peacock also have a significant place in our psycho-spiritual arena.

(iv) Ethical Value

a) It is also sometimes known as existence value. It involves ethical issues like "all life must be preserved".
b) The ethical value means that we may or may not use a species, but knowing the very fact that this species exists in nature gives us pleasure.

c) We are not deriving anything direct from Kangaroo, Zebra or Giraffe, but we all strongly feel that these species should exist in nature.

(v) Aesthetic Value

a) No one of us would like to visit vast stretches of barren lands with no signs of visible life.

b) People from far and wide spend a lot of time and money to visit wilderness areas where they can enjoy the aesthetic value of biodiversity and this type of tourism is now known as eco-tourism.

c) Ecotourism is estimated to generate about 12 billion dollars of revenue annually.

(vi) Option Values

a) These values include the potentials of biodiversity that are presently unknown and need to be explored.

b) There is a possibility that we may have some potential cure for AIDS or cancer existing within the depths of a marine ecosystem, or a tropical rain forest.

c) Thus option value is the value of knowing that there are biological resources existing on this biosphere that may one day prove to be an effective option for something important in the future.

(vii) Ecosystem Service Value

a) It refers to the services provided by ecosystems like prevention of soil erosion, prevention of floods, maintenance of soil fertility, cycling of nutrients, prevention floods, cycling of water, their role as carbon sinks, pollutant absorption, and reduction of the threat of global warming etc.

India as Mega Diversity Nation

India is one of the 12 megadiversity countries in the world. The Ministry of Environment and forests, Govt. of India (2000) records 47,000 species of plants and 81,000 species of animals which are about 7% and 6.5% respectively of global flora and fauna.

Endemism: Species, which are restricted only to a particular area, are known as endemic. India shows a good number of endemic species. About 62% of amphibians and 50% of lizards is endemic to India.

Center of origin: A large number of species are known to have originated in India. Nearly 5000 species of flowering plants had their origin in India.

Marine diversity: Along 7500 km long coastline of our country in the mangroves, estuaries, coral reefs, backwaters etc. There exists a rich biodiversity. More than 340 species of corals of the world are found here.

A large proportion of the Indian Biodiversity is still unexplored. There are about 93 major wetlands, coral reefs, and mangroves which need to be studied in detail.

Hotspots of Biodiversity

- Areas, which exhibit high species richness as well as high species endemism, are termed as hot spots of biodiversity.
- The term was introduced by Myers (1988).
- There are 25 such hot spots of biodiversity on a global level out of which two are present in India, namely the Eastern Himalayas and the Western Ghats.
- These hot spots covering less than 2% of the world's land are found to have about 50% of the terrestrial biodiversity.
- About 40% of terrestrial plants and 25% of vertebrate species are endemic and found in these hotspots.
- After the tropical rain forests, the second highest number of endemic plant species are found in the Mediterranean (Mittermeier).
- Earlier 12 hot spots were identified on a global level.
- Later Myers et al (2000) recognized 25 hot spots.
- Two of these hotspots lie in India extending into neighbouring countries namely, Indo-Burma region (covering Eastern Himalayas) and the Western Ghats – Sri Lanka region.
- The Indian hot spots are not only rich in floral wealth and endemic species of plants but also reptiles, amphibians, swallow-tailed butterflies and some mammals.

(a) Eastern Himalayas

They display an ultra-varied topography that fosters species diversity and endemism. Certain species like Sapria Himalayan, a parasitic angiosperm was sighted only twice in this region in the last 70 years. Out of the world's recorded flora, 30% are endemic to India of which 35,000 are in the Himalayas.

(b) Western Ghats

It extends along a 17,000 Km² strip of forests in Maharashtra, Karnataka, Tamil Nadu and Kerala and has 40% of the total endemic plant species. 62% amphibians and 50% lizards are endemic to the Western Ghats. The major centers of diversity are Agastyamalai Hills and Silent Valley-the New Amambalam reserve Basin. It is reported that only 6.8% of the original forests are existing today while the rest has been deforested or degraded. Although the hotspots are characterized by endemism, interestingly, a few species are common to both the hotspots in India.

Threats to Biodiversity

Extinction or elimination of a species is a natural process of evolution. In the geologic period, the earth has experienced mass extinctions. During evolution, species have died out and have been replaced by others. The process of extinction has become particularly fast in the recent years of human civilization. One of the estimates by the noted ecologist, E.O. Wilson puts the figure of extinction at 10,000 species per year or 27 per day! This startling figure raises an alarm regarding the serious threat to biodiversity.

Let us consider some of the major causes and issues related to threats to biodiversity.

1) **Loss of Habitat**
   - Destruction and loss of natural habitat are the single largest cause of biodiversity loss. Billions of hectares of forests and grasslands have been cleared over the past 10,000 years for conversion into agriculture lands, pastures, settlement areas or development projects.
   - There has been a rapid disappearance of tropical forests in our country also, at a rate of about 0.6% per year.
   - With the current rate of loss of forest habitat, it is estimated that 20-25% of the global flora would be lost within a few years.
   - Marine biodiversity is also under serious threat due to large scale destruction of the fragile breeding and feeding grounds of our oceanic fish and other species, as a result of human intervention.

2) **Poaching**
   - Illegal trade of wildlife products by killing prohibited endangered animals i.e. poaching is another threat to wildlife.
   - Despite an international ban on trade in products from endangered species, smuggling of wildlife items like furs, hides, horns, tusks, live specimens and herbal products worth millions of dollars per year continues.
   - The cost of elephant tusks can go up to $100 per kg; the leopard fur coat is sold at $ 100,000 in Japan while bird catchers can fetch up to $ 10,000 for a rare hyacinth macaw, a beautiful colored bird, from Brazil.

3) **Man-Wildlife Conflict**
   - Instances of man-animal conflicts keep on coming to limelight from several states in our country.
   - In Sambalpur, Orissa 195 humans were killed in the last 5 years by elephants.

- In retaliation, the villagers killed 98 elephants and badly injured 30 elephants.
- Several instances of killing of elephants in the border regions of Kote-Chamarajanagar belt in Mysore have been reported recently.
- The man-elephant conflict in this region has arisen because of the massive damage was done by the elephants to the farmer's cotton and sugarcane crops.
- The agonized villagers electrocute the elephants and sometimes hide explosives in the sugarcane fields, which explode as the elephants intrude into their fields.
- In early 2004, a man-eating tiger was reported to kill 16 Nepalese people and one 4-year old child inside the Royal Chitwan National Park of Kathmandu.
- In June 2004, two men were killed by the leopards in Powai, Mumbai.

Cause of Man-Animal Conflicts

1) Dwindling habitats of tigers, elephants, and bears due to shrinking forest cover compels them to move outside the forest and attack the field or sometimes even humans.

2) Usually, the ill, weak and injured animals have a tendency to attack the man. Also, the female tigress attacks the human if she feels that her newborn cubs are in danger. But the biggest problem is that if human-flesh is tasted once then the tiger does not eat any other animal.

3) Earlier, forest departments used to cultivate paddy, sugarcane etc. within the sanctuaries when the favorite staple food of elephants i.e. bamboo leaves was not available. Now due to lack of such practices the animals move out of the forest in search of food.

4) Very often the villagers put electric wiring around their ripe crop fields. The elephants get injured, suffer in pain and turn violent.

5) The cash compensation paid by the government in lieu of the damage caused to the farmer's crop is not enough. The agonized farmer, therefore, gets revengeful and kills the wild animals.

Remedial Measures to Curb the Conflict

1) Tiger Conservation Project (TCP) has made provisions for making available vehicles, tranquillizer guns, and binoculars to tactfully deal with any imminent danger.

2) Adequate crop compensation and cattle compensation scheme must be started.

3) Solar powered fencing should be provided along with electric current proof trenches to prevent the animals from straying fields.

4) The cropping pattern should be changed near forest borders and adequate fruits and water should be made available for the elephants within forest zones.

5) Wildlife corridors should be provided for mass migration of big animals during unfavorable periods.

Endangered Species of India

- The International Union for Conservation of Nature and Natural Resources (IUCN) publishes the Red Data Book which includes the list of endangered species of plants and animals.

- The **red data** symbolizes the warning signal for those species which are endangered and if not protected are likely to become extinct in near future.

- In India, nearly 450 plant species have been identified in the categories of endangered, threatened or rare.

- The existence of about 150 mammals and 150 species of birds is estimated to be threatened while an unknown number of species of insects are endangered.

- A few species of endangered reptiles, birds, mammals and plants are given below.
    - **Reptiles:** Green sea turtle, tortoise, python
    - **Birds:** Great Indian bustard, Peacock, Pelican, Great Indian Hornbill, Siberian
    - **Carnivorous Mammals:** Indian wolf, red fox, red panda, tiger, leopard, Indian, lion, golden cat, desert cat
    - **Primates:** Hoolock gibbon, capped monkey, golden monkey
    - **Plants:** A large number of species of orchids, Rhododendrons, medicinal plants like Rauvolfia serpentina, the sandal, wood tree santalum, cycas beddonei etc

The Zoological Survey of India reported that Cheetah, Pink-headed duck and mountain quail have already become extinct from India.

- A species is said to extinct when it is not seen in the wild for 50 years at a stretch eg. Dodo, Passenger pigeon.

- A species is said to be endangered when its number has been reduced to a critical level. If such a species is not protected and conserved, it is in immediate danger of extinction.

- A species is said to be invulnerable category if its population is facing a continuous decline due to overexploitation or habitat destruction.

- Species which are not endangered or vulnerable at present, but are at a risk are categorized as rare species.

Plate IV. Some important extinct and endangered Indian species of animals.

Endemic Species

- The species are only found among a particular people or in a particular region are known as endemic species.
- Out of about 47, 00 species of plants in our country 7000 are endemic.
- Some of the important endemic flora includes orchids and species like sapria himalayana, Uvaria lureda, Nepenthes khazi ana etc.
- A large number out of total 81,000 species of animals in our country is endemic.
- The Western Ghats are particularly rich in amphibians and reptiles.
- About 62%amphiians and 50% lizards are endemic to the Western Ghats.
- Different species of monitor lizard, reticulated python are some important endemic species of our country.

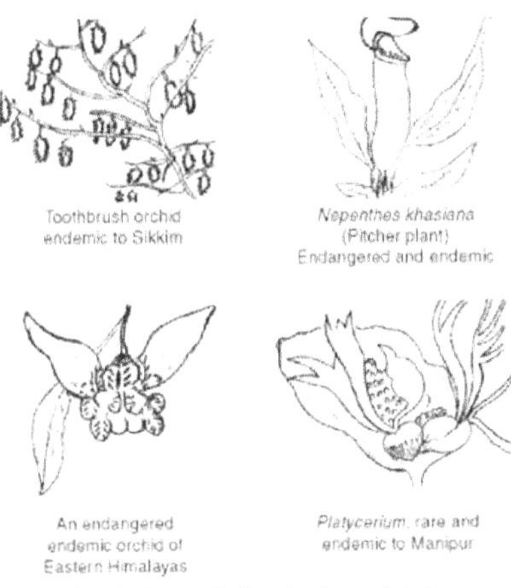

Plate V. Some endemic and endangered plants

Conservation of Biodiversity

The enormous value of biodiversity due to their genetic, commercial, medical, esthetic, ecological and optional importance emphasizes the need to conserve biodiversity.

There are two approaches to biodiversity conservation:

a) **In situ conservation (within habitat):** This is achieved by protection of wild flora and fauna in nature itself. E.g. Biosphere Reserves, National Parks, Sanctuaries, Reserve Forests etc.

b) **Ex-situ conservation (outside habitats):** This is done by the establishment of gene banks, seed banks, zoos, botanical gardens, culture collections etc.

In Situ Conservation

At present in our country we have:

- 7 major Biosphere reserves.
- 80 National Parks.
- 420 wildlife sanctuaries.
- 120 Botanical gardens.
- They totally cover 4% of the geographic area.

The Biosphere Reserves conserve some representative ecosystems as a whole for long-term in situ conservation.

In India we have:

- Nanda Devi (U.P.).
- Nokrek (Meghalaya).
- Manas (Assam).
- Sunderbans (West Bengal).
- Gulf of Mannar (Tamil Nadu).
- Nilgiri (Karnataka, Kerala, Tamil Nadu).
- Great Nicobars and Similipal (Orissa).

A National Park is an area dedicated to the conservation of wildlife along with its environment. It is also meant for enjoyment through tourism but without impairing the environment. Grazing of domestic animals, all private rights and forestry activities are prohibited within a National Park. Each National Park usually aims at conservation specifically of some particular species of wildlife along with others.

Some important National parks in India.

Name of National Park	State	Important Wildlife
Kaziranga	Assam	One horned Rhino
Gir National Park	Gujarat	Indian Lion
Bandipur	Karnataka	Elephant
Periyar	Kerala	Elephant, Tiger
Sariska	Rajasthan	Tiger

Wildlife sanctuaries are also protected areas where killing, hunting, shooting or capturing of wildlife is prohibited except under the control of the highest authority. Some major wildlife sanctuaries of our country are shown in table.

Some Important Wildlife Sanctuaries of India.

Name of Sanctuary	State	Major Wild Life
Ghana Bird Sanctuary	Rajasthan	3oo species of birds (including migratory)
Sultanpur Bird Sanctuary	Haryana	Migratory birds
Mudumalai Wildlife Sanctuary	Tamil Nadu	Tiger, elephant, Leopard
Vedanthangal Bird Sanctuary	Tamil Nadu	Water birds
Wild Ass Sanctuary	Gujarat	Wild ass, wolf, nilgai

For plants, there is one gene sanctuary for Citrus (Lemon family) and one for pitcher plant (an insect eating plant) in Northeast India.

Ex Situ Conservation

This type of conservation is mainly done for conservation of crop varieties. In India, we have the following important gene bank/seed bank facilities:

National Bureau of Plant Genetic Resources (NBPGR) is located in New Delhi. Here agricultural and horticultural crops and their wild relatives are preserved by cryopreservation of seeds, pollen etc. by using liquid nitrogen at a temperature as low as – 196 degree Celsius. Varieties of rice, turnip, radish, tomato, onion, carrot, chilli, tobacco etc. have been preserved successfully in liquid nitrogen for several years without losing seed viability.

National Bureau of Animal Genetic Resources (NBAGR) located at Karnal, Haryana. It preserves the semen of domesticated bovine animals.

National Facility for Plant Tissue Culture Repository (NFPTCR) for the development of a facility of conservation of varieties of crop plants/trees by tissue culture. This facility has been created within the NBPGR. For the protection and conservation of certain animals, there have been specific projects in our country e.g. Project Tiger, Girl Lion Project, Crocodile Breeding Project, Project Elephant, Snow Leopard Project etc.

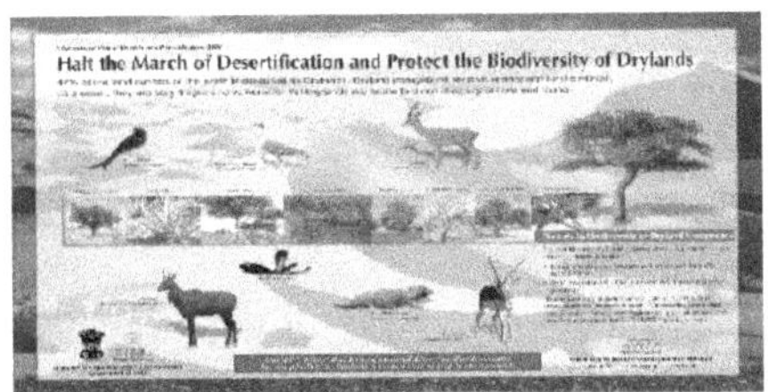

International day for Biological Diversity - 22nd May

A field study of common plants, insects, birds. A field study of simple ecosystems – pond, river, hill, slopes.

UNIT III

ENVIRONMENTAL POLLUTION

Environmental pollution is defined as any undesirable change in the physical, chemical or biological characteristics of any component of the environment (air, water, soil) which can cause harmful effects on various forms of life and property.

3.1. Pollutant

Any trace of element responsible for creating pollution is known as pollutants

Classification of Pollutants

- Biodegradable pollutants - decompose rapidly by natural processes.
- Non- degradable pollutants-do not decompose or slowly decompose in the environment.

Types of Pollution

1. Air pollution.
2. Water pollution.
3. Soil pollution.
4. Marine pollution.
5. Noise pollution.
6. Thermal pollution.
7. Nuclear hazards.

3.2.   Air Pollution

Air pollution may be defined as, the presence of one or more contaminants like dust, smoke, mist and odour in the atmosphere which are injurious to human beings, plants, and animals.

Causes

- Rapid industrialization.
- Fast urbanization.
- Rapid growth of population.
- Drastic increase in vehicles.
- Human activities.

Sources of Pollutants

Gaseous pollutants – sulphur oxides, nitrogen oxides, carbon oxides, volatile organic compounds

Particulate pollutants – smoke, dust, soot, fumes, aerosol, liquid droplets, pollen grains

Radioactive pollutants – Radon 222, Iodine 131, Sr 90

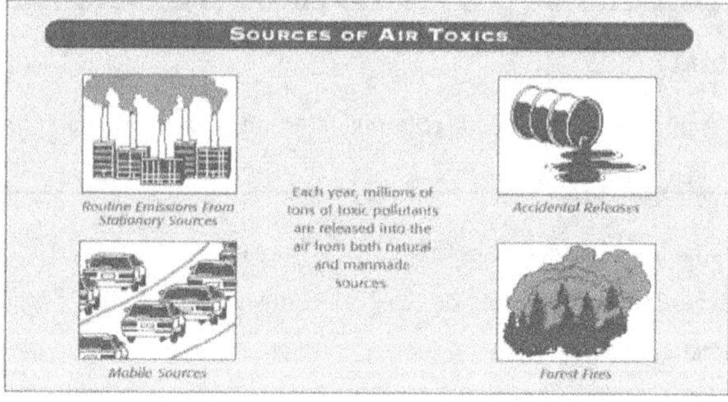

Sources of Air Pollution

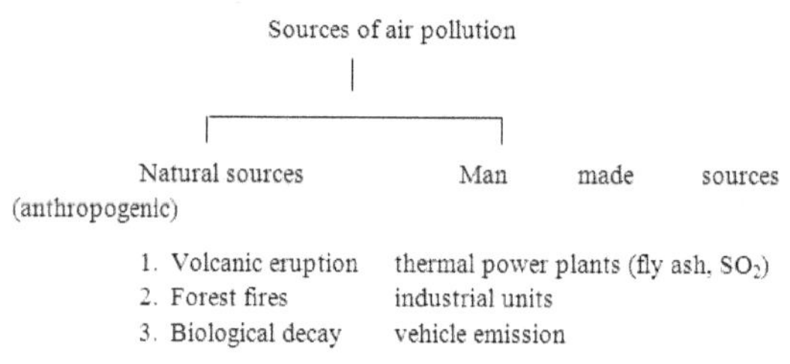

Classification of Air Pollutants

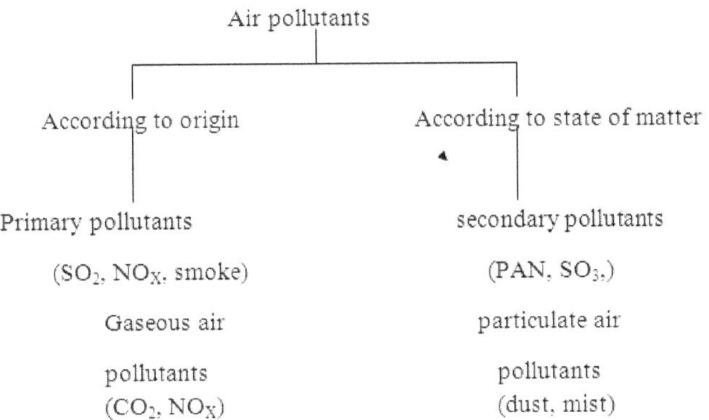

Primary Air Pollutants

Primary air pollutants are those emitted directly in the harmful form. Eg. CO, NO, SO_2 etc.

1. Indoor Air Pollution

Radon is an important air pollutant. It can be emitted from building materials like bricks, concrete, tiles etc. which are derived from soil containing radium. Burning of fuel produce pollutants like CO, SO_2, soot and many other like formaldehyde, benzo (a) pyrene (BAP) are toxic and harmful for health. BAP is also found in cigarette smoke and is considered to cause cancer. A person using wood as fuel for cooking inhales BAP equivalent to 20 packets of cigarette a day.

Secondary Air Pollutants

Some of the primary air pollutants react with one another or with the basic components of air to form new pollutants. They are called secondary pollutants.

Common Pollutants, Sources and their Effects

According to WHO (world health organization), more than 1.1billion people live in an urban area where outdoor air is unhealthy to breathe. Some of the common air pollutants are described below.

Pollutants	Description	Sources/causes	Human effect/ environmental effect
Carbon monoxide (CO)	Colorless, odorless gas. It is formed by partial combustion of carbon-containing fuels. $2C+O_2 \rightarrow 2CO$	Cigarette smoking, incomplete burning of fossil fuels.	**H.E:** Reacts with haemoglobin in red blood cells and reduces the ability of blood to bring oxygen to body cells and tissues which cause headaches and anemia. **E.E:** increases global temperature
Nitrogen dioxide(NO_2)	Reddish brown gas that gives photo -chemical smog. It reacts with atmospheric moisture and converted into nitric acid.$NO_2+H_2O \rightarrow HNO_3$	Fossil fuels burning in motor vehicles (49%) and power industrial plant (49%)	**H.E:** lung irritation and damage **E.E:** acid deposition can damage trees, soils, aquatic life and also corrode metals (monuments, buildings, and statues)
Sulphur Dioxide (SO_2)	Colourless and irritating gas. It is formed by the combustion of fuels like coal and oil. It is converted into sulphuric acid while reacting with atmospheric moisture.	Coal burning power plants (88%) and industrial processes (10%)	**H.E:** lung irritation and damage **E.E:** acid deposition can damage trees, soils, aquatic life and also corrode metals (monuments, buildings, and statues)
Suspended particulate matter (SPM)	It includes varieties of particles and droplets (aerosols).	Burning of coal in power and industrial plants (40%), burning diesel (17%)and construction	**H.E:** Nose and throat irritation, lung damage, bronchitis, asthma, and cancer. **E.E:** Reduces visibility. acid deposition can damage trees, soils, aquatic life
Ozone	It is a major component of photochemical smog. Highly reactive irritating gas with an unpleasant odour.	A chemical reaction with volatile organic compounds and nitrogen oxides.	Moderate the climate.
Photochemical smog	Brownish smoke like appearance that frequently forms on clear, sunny days over large cities with significant automobile traffic	Chemical reactions among nitrogen oxides and hydrocarbon by sunlight.	**H.E:** Breathing problem, cough, heart diseases eye, nose and throat irritation. **E.E:** It damages plants and trees. Smog can reduce visibility.
Lead and chromium	Solid toxic metal and its compounds emitted into the atmosphere as particulate matter.	Paint, smelters (metal refineries) lead manufacture, storage batteries, leaded petrol	**H.E:** Accumulates in the body, brain and other nervous system damage and mental retardation, digestive and other health problems **E.E:** can harm wildlife.
Hydrocarbons	Hydrocarbons especially lower hydrocarbons get accumulated due to the decay of vegetable matter	Agriculture, the decay of plants, burning of wet logs.	**H.E:** carcinogenic **E.E:** It produces an oily film on the surface and does not as such cause a serious problem until they react to form secondary pollutants. Ethylene causes plant damage even at low concentration.

Control of Air Pollution

1. Source Control

- Use only unleaded petrol.
- Use petroleum products and other fuels that have low sulphur and ash contents.
- Reduce the number of private vehicles on the road by developing an efficient public transport system and encouraging people to walk or use cycles.
- Ensure that houses, schools, restaurants and places where children play are not located on busy streets.
- Plant trees along busy streets because they remove particulates and carbon monoxide, and absorb noise.
- Industries and waste disposal sites should be situated outside the city centre preferably downwind of the city.
- Use catalytic converters to help control emissions of carbon monoxide and hydrocarbons.

Reduction of Air Pollution at Industrial Centers

- The emission rates should be restricted to permissible levels by each and every industry.
- Incorporation of air pollution control equipment in the design of the plant layout must be made mandatory.

- Continuous monitoring of the atmosphere for the pollutants should be carried out to know the emission levels.

Many devices are available nowadays, but the choice of which depends on characteristics, flow rate, cost, efficiency.

Cyclone Separator

The general principle of inertia separation is that the particulate-laden gas is forced to change direction. As gas changes direction, the inertia of the particles causes them to continue in the original direction and be separated from the gas stream. The walls of the cyclone narrow toward the bottom of the unit, allowing the particles to be collected in a hopper. The cleaner air leaves the cyclone through the top of the chamber, flowing upward in a spiral vortex, formed within a downward moving spiral. Cyclones are efficient in removing large particles but are not as efficient with smaller particles. For this reason, they are used with other particulate control

Bag House Filters

It consists of several filter bags made up of fabric which is hung upside down in several compartments. Dirty gas is passed through this bag house which leaves the bags through the pores. The dust particles settled in the form of cake is removed by shaking. This method is efficient for small particles and is used in industries. It is more expensive as it can't be used for moist gases, corrosive gases. Depending upon the nature of the elements present, it can be chosen.

Wet Scrubbers

Dirty gas is passed through the water or water is sprayed on the gas. As the particles become wet it from the top of the scrubber. This method is very efficient for removing the particulates and also for the removal of toxic and acidic gases.

Electrostatic Precipitators

It may be a plate type or cylinder type. Vertical wires are placed between the parallel plates or wire is hung along the axis of the cylinder. High negative voltage is applied to the wire. Dust particles while passing from the lower end gets negatively charged and are collected at the positive surface while the clean gas leaves from the top. The deposited particles are removed by scrubbing. This method involves the removal of even submicroscopic particles.

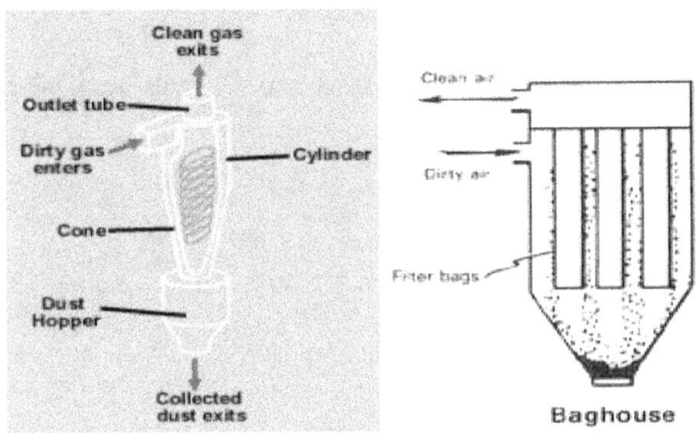

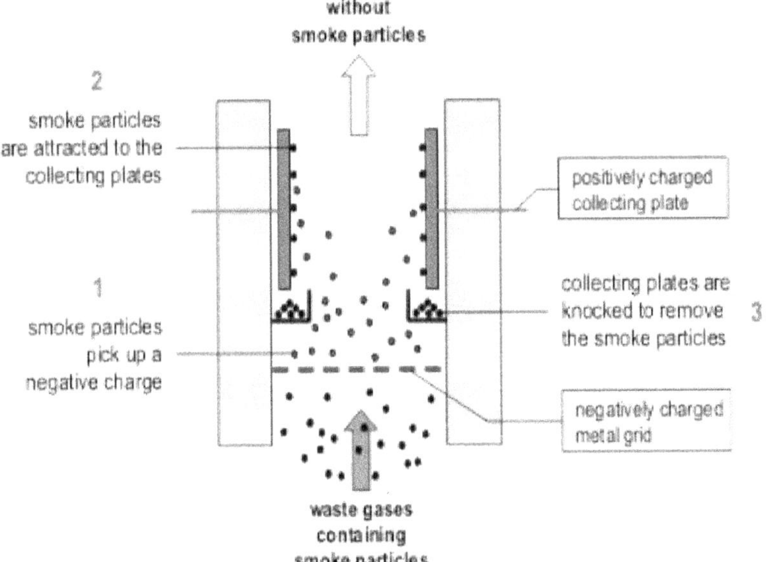

Global Warming

Global warming is the increase in the average temperature of the Earth's near-surface air and oceans since the mid-20th century and its projected continuation. Global surface temperature increased 0.74 ± 0.18 °C (1.33 ± 0.32 °F) during the last century. The Intergovernmental Panel on Climate Change (IPCC) concludes that increasing greenhouse gas concentrations resulting from human activity such as fossil fuel burning and deforestation caused most of the observed temperature increase since the middle of the 20th century.

Green House Effect

The progressive warming up of the earth's surface due to blanketing effect of manmade CO2 in the atmosphere.

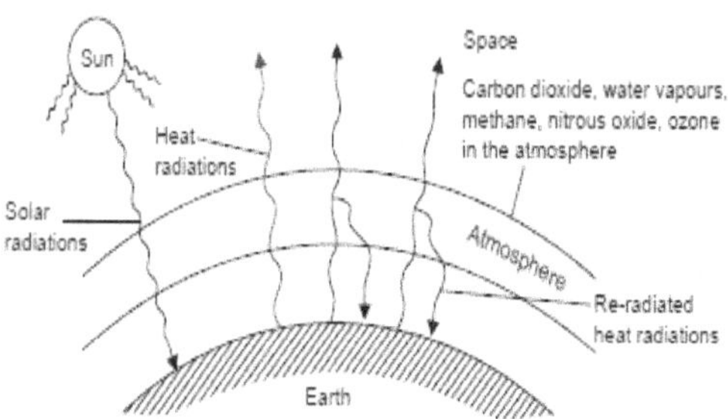

Causes of Global Warming

1. *Carbon Dioxide*

The main source of fossil fuel burning (67%) and deforestation, other forms of land clearing and burning (33%).CO2 concentration in the atmosphere was 355 ppm in 1990 that is increasing at a rate of 1.5 ppm every year.

2. *Chlorofluorocarbons*

The main sources of CFC's include leaking air conditioners and refrigerators, evaporation of industrial solvents, production of plastic foams, aerosols, propellants etc. The atmospheric concentration of CFC is 0.00225 ppm that is increasing at a rate of 0.5% annually.

3. *Methane*

While carbon dioxide is the principal greenhouse gas, methane is second most important. According to the IPCC, Methane is more than 20 times as effective as CO2 at trapping heat in the atmosphere.

4. *Nitrous Oxide*

Another greenhouse gas is a Nitrous oxide (N2O), a colourless, non-flammable gas with a sweetish odour, commonly known as "laughing gas", and sometimes used as an anesthetic. Nitrous oxide is naturally produced by oceans and rainforests. Man-made sources of nitrous oxide include nylon and nitric acid production, the use of fertilizers in agriculture, cars with catalytic converters and the burning of organic matter.

Effects of Global Warming

1. Effect on sea level – As a result of the glacial melting thermal expansion of the ocean, a 20 cm rise is expected in sea level by 2030.
2. Effect on agriculture and forestry – High CO2 level in the atmosphere has long-term negative effects on crop production and forest growth.
3. Effect on water resources – Global rainfall patterns will change and the water management strategies of different regions will need to adapt to these changes.
4. Effect on terrestrial ecosystems – Many species will be at risk from extinction whereas more tolerant varieties will thrive.
5. Effect on human health – Three would be increased in waterborne diseases, infectious diseases carried by mosquitoes and other diseases vectors.

Preventive Measures

- CO_2 emission can be cut by reducing the use of fossil fuels.
- Implement energy conservation measures.
- Utilize renewable resources such as wind, solar and hydropower.
- Plant more trees.
- The shift from coal to natural gas.
- Adopt sustainable agriculture.
- Stabilize population growth.
- Efficiently remove CO_2 from smoke stacks.
- Remove atmosphere CO_2 by utilizing photosynthetic algae.

Ozone Layer Depletion

Ozone is a gas($O3$) found in the atmosphere, but most highly concentrated in the stratosphere between 10 and 50 km above sea level, where it is known as the ozone layer.

Causes for Ozone Depletion

Chloroflouro carbons (CFC) commonly known as Freon's. These chemically stable (non-reactive), odorless, nonflammable, nontoxic and non-corrosive compounds are cheap to make. They are used as coolants in air conditioners and refrigerators (replacing toxic sulfur dioxide and ammonia), propellants in an aerosol spray can, Cleaners for electronic parts, sterility for hospital instruments, Fumigants of granaries and bubbles in plastic foam used for Insulation and packaging.

Ozone Depletion in Stratosphere

A layer of ozone in the lower stratosphere keeps about 95% of the sun's harmful UV radiation from reaching the earth's surface.

Formation of Ozone

Ozone is a form of oxygen and it contains three atoms of oxygen (O3) in the stratosphere .ozone is continuously being created by the absorption of short wave UV radiations. UV radiations less than 242nm decompose molecular oxygen into atomic (O) by photolytic decomposition.

$$(O2+h\gamma \text{----} O\bullet+O\bullet)$$

The atomic oxygen rapidly reacts with molecular oxygen to form ozone.

$$(O2+O\bullet \text{----} O3)$$

Some free oxygen atoms combine with O3 to form two oxygen molecules.

$$O+ O3 \text{--------} O2 + O2$$

When ozone absorbs UVB, it is converted to free oxygen and molecular oxygen.

$$O3 + UVB \text{------} O + O2$$

The amount of ozone in the stratosphere is dynamic. However, the presence of CFC and other chemicals may alter this equilibrium. Ozone Concentration in the stratosphere is about 10ppm. This is disturbed by reactive atoms of chlorine, bromine etc, which destroy ozone, molecules. It results in thinning of ozone layer generally called ozone hole. The amount of One DU =0 .01 mm.

Mechanism of Ozone Layer Depletion

In 1974, Chemists Sherwood Rowland and Mario Molina at the University of California indicated that CFCS were lowering the average concentration of ozone in the stratosphere. CFCs remain the troposphere because they are insoluble in water and are chemically un reactive. Over, 11-20 years, they rise into the stratosphere mostly through convection, random drift and the turbulent mixing of air in the troposphere. Once they reach the stratosphere, the CFC, molecules break down under the influence of high-energy UV radiation. It releases highly reactive ozone (O3) into O2 and O in a cyclic chain of chemical reactions. This causes ozone in various parts of the stratosphere to be destroyed faster than it is formed.

$$CF2Cl2 + UV \quad CF2Cl + Cl$$
$$CF2Cl+O2 \rightarrow CF2O+ClO$$

$$Cl + O3 ---- ClO + O2$$
$$ClO + O\bullet ----- Cl + O2$$

Finally, each CFC can last in the stratosphere for 65 -385 years. Depending on its type – the most widely used CFCs last 75-111 years. During that time, each chlorine atom released from these molecules can convert unto 1,00,000 molecules of O3 to O2. Other Ozone depleting substances CFCs are not the only ozone depleting compounds. Other examples are halons.

Hydro Chloro Fluoro Carbon, Bromo Fluoro Carbons, Methyl bromide (fumigant), hydrogen chloride (emitted by US space shuttles) and cleaning solvents such as carbon tetrachloride, methyl chloroform, n-propyl bromide and hexachloride butadiene.

Effects of Ozone Layer Depletion

Human Health

- The UV rays damage genetic material in the skin cells which cause skin cancer.
- Increases the risk of nonMalinine skin cancer.
- Slow blindness called actinic keratitis. The enhanced level could lead to more people suffering from cataracts.
- Suppress the immune responses in humans and animals and reduces human resistivity leading to a number of diseases such as cancer, allergies, and other infectious diseases.

Aquatic Systems

- Affect the aquatic forms such as phytoplankton, fish, larval crabs.
- The phytoplankton consumes large amount of CO2.

Materials

- Degradation of paints, plastics, and other polymeric material will result in an economic loss due to effects of UV radiation.

Climate

- Increases the average temperature of the earth's surface.

Acid Rain

Acid rain is rain or any other form of precipitation that is unusually acidic. It has harmful effects on plants, aquatic animals, and infrastructure. Acid rain is mostly caused by human emissions of sulfur and nitrogen compounds which react in the atmosphere to produce acids.

Definition

"Acid rain" is a popular term referring to the deposition of wet (rain, snow, sleet, fog and cloud water, dew) and dry (acidifying particles and gases) acidic components. A more accurate term is"acid deposition". Distilled water, which contains no carbon dioxide, has a neutral pH of 7. Liquids with a pH less than 7 are acidic, and those with a pH greater than 7 are bases. "Clean" or unpolluted rain has a slightly acidic pH of about 5.2 because carbon dioxide and water in the air react together to form carbonic acid, a weak acid (pH 5.6 in distilled water), but unpolluted rain also contains other chemicals.

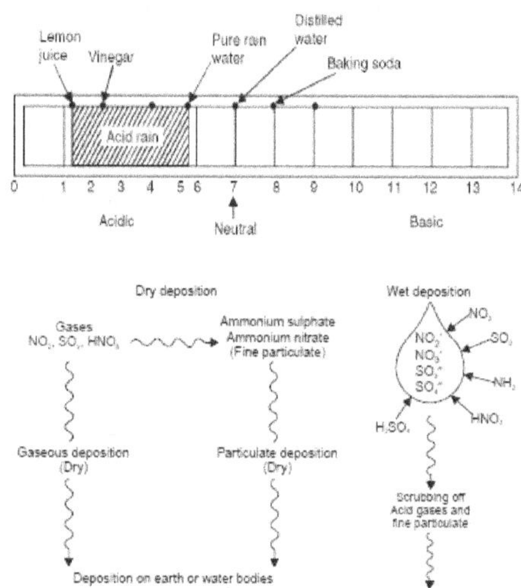

Acid Deposition (dry Deposition and Wet Deposition)

$$H2O\ (l) + CO2\ (g) \rightarrow H2CO3\ (aq)$$

Carbonic acid then can ionize in water forming low concentrations of hydronium and carbonate ions:

$$2\ H_2O\ (l) + H_2CO_3\ (aq) \quad CO_3{}^{2-}\ (aq) + 2\ H_3O^+\ (aq)$$

Emissions of Chemicals Leading to Acidification

The most important gas which leads to acidification is sulfur dioxide. Emissions of nitrogen oxides which are oxidized to form nitric acid are of increasing importance due to stricter controls on emissions of sulfur containing compounds. 70 Tg(S) per year in the form of SO2 comes from fossil fuel combustion and industry, 2.8 Tg(S) from wildfires and 7-8 Tg(S) per year from volcanoes.

Natural Phenomena

The principal natural phenomena that contribute acid-producing gases to the atmosphere are emissions from volcanoes and those from biological processes that occur on the land, in wetlands, and in the oceans. The major biological source of sulfur-containing compounds is dimethyl sulfide.

Human Activity

The principal cause of acid rain is sulfur and nitrogen compounds from human sources, such as electricity generation, factories, and motor vehicles. Coal power plants are one of the most polluting. The gases can be carried hundreds of kilometers in the atmosphere before they are converted to acids and deposited.

Chemical Processes

Combustion of fuels creates sulfur dioxide and nitric oxides. They are converted into sulfuric acid and nitric acid.

Gas Phase Chemistry

In the gas phase, sulfur dioxide is oxidized by reaction with the hydroxyl radical via an intermolecular reaction.

$$SO_2 + OH\cdot \rightarrow HOSO_2\cdot$$

Which is followed by

$$HOSO_2\cdot + O_2 \rightarrow HO_2\cdot + SO_3$$

In the presence of water, sulfur trioxide (SO3) is converted rapidly to sulfuric acid:

$$SO_3\ (g) + H_2O\ (l) \rightarrow H2SO_4\ (l)$$

Nitrogen dioxide reacts with O to form nitric acid:

$$NO2 + OH\cdot \rightarrow HNO_3$$

Wet Deposition

Wet deposition of acids occurs when any form of precipitation (rain, snow, etc.) removes acids from the atmosphere and delivers it to the Earth's surface. This can result from the deposition of acids produced in the raindrops (see aqueous phase chemistry above) or by the precipitation removing the acids either in clouds or below clouds. Wet removal of both gases and aerosols are both of importance for wet deposition.

Dry Deposition

Acid deposition also occurs via dry deposition in the absence of precipitation. This can be responsible for as much as 20 to 60% of total acid deposition. This occurs when particles and gases stick to the ground, plants or other surfaces.

Adverse Effects

- Effect of acid rain on human beings – Human respiratory system, nervous system, and digestive system are affected by acid rain, and causes the premature death from heart and lung disorder such as asthma and bronchitis.
- Effect of acid rain on buildings – The TajMahal in Agra suffering at present due to SO2 and H2SO4 acid fumes released from Mathura refinery. Acid rain corrodes houses, monuments, statues, bridges and fences. British parliament building also suffered damage due to H2SO4 rains. Dry deposition of acidic compounds can also dirty buildings and other structures, leading to increased maintenance costs.
- Effect on Terrestrial and lake ecosystem – reduces the rate of photosynthesis and growth and increased sensitivity to drought and disease, retards the growth of crops such as beans, radish, potato, spinach and carrots. The activity of bacteria and other microscopic animals is reduced in acidic water

Case Study

The Bhopal Gas Tragedy

The world's worst industrial accident occurred in Bhopal, M.P., India on the night of 2nd and morning of 3rd December 1984. It happened at Union Carbide Company which used to manufacture Carbaryl (Carbamate) pesticide using Methyl isocyanate (MIC). Due to the accidental entry of water in the tank, the reaction mixture got overheated and exploded because its cooling system had failed. Other safety devices also did not work or were not in the working condition. Forty tons of MIC leaked into the atmosphere which might have contained 40 kg of phosgene as an impurity. MIC gas at lower concentrations affects lungs and eyes and causes irritation in the skin. Higher amounts remove oxygen from the lungs and can cause death. In the winter night of December, there was fog like clouds over south and east of the plant. The gas spread over 40 Km^2 Area. About 5100 persons were killed (2600 due to direct exposure to MIC and other 2500 due to after effects of exposure) according to Indian officials. About 2,50,000 persons got exposed to MIC. An estimated 65,000 people suffered from the severe eye, respiratory, neuromuscular, gastrointestinal and gynecological disorders. About 1000 persons became blind. Without counting the damage of human lives, it cost about $ 570 million in cleanup and damage settlement. This tragedy could have been averted had the company spend about $ 1 million on safety improvement

3.3. Water Pollution

It may be defined as "the alteration in physical, chemical and biological characteristics of water which may cause harmful effects on human and aquatic life.

Types

Point source: point sources are discharged pollutants at specific locations through pipes, ditches or sewers into bodies of surface water. **ex**: the flow of water pollutants from sewerage system, industrial effluent etc.

Non-point source: They are usually large land areas or airsheds that pollute water by runoff, subsurface flow or deposition from the atmosphere.

Location of which cannot be easily identified. **Ex**: agricultural land (pesticides, fertilizers, mining, construction sites)

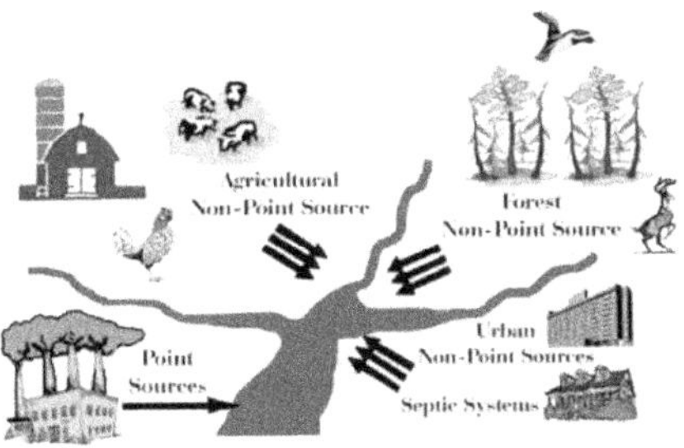

Pollutants	Source/cause	Effects
Infectious agents	Human and animal waste eg. Bacteria, viruses,protozoa, and parasitic worms	Variety of diseases
Inorganic chemicals	Surface runoff, industrial effluents and household cleansers. Eg. Acids, toxic metals such as lead, arsenic, and selenium, salts like chlorides and fluorides.	(i)Skin cancer and neck damage (ii)Damage the nervous system, liver and kidney (iii)Harm fish and other aquatic life (iv)Lower crop yields
Organic chemicals	Industrial effluents, household cleaner, surface runoff from farms. Eg. Oil, gasoline, plastics, pesticides, cleaning solvents and detergents	(i) can threaten human health by causing nervous system damage and some cancers. (ii) harm fish and wildlife
Plant nutrients	Sewage, manure, and runoff of agricultural and urban fertilizers. Eg. Water –soluble compounds containing nitrate, phosphate and ammonium ions	(i) Can cause excessive growth of algae and other aquatic plants, which die, decay, deplete dissolved oxygen in the water and kill the fish. (ii)drinking water with excessive levels of nitrates lower the oxygen-carrying capacity of the blood and can kill urban children and infants
Radioactive materials	Nuclear power plants, mining, and processing of uranium and other ores, nuclear weapons production and natural sources. Eg. Radioactive isotopes of iodine, radon, uranium, cesium and thorium	Genetic mutations, birth defects and certain cancers.
Sediments	Land erosion eg. Soil, silt etc	(i)Can reduce photosynthesis and cloud water. (ii)Disrupt aquatic food webs. (iii)Carry pesticides ,bacteria and other harmful substance (iv)settle out and destroy feeding and spawning grounds of fish (v) Clog and fill lakes artificial reservoirs, stream channels, and harbors.
Thermal pollutants	Water cooling of electric power plants and some types of industrial plants. Eg. Excessive heat	Lowers the dissolved oxygen levels and makes aquatic organisms more vulnerable to disease.

(Note: When power plant first opens or shut down for repair, fish and other organisms adapted to a particular temperature range can be killed by the abrupt change in water temperature known as thermal shock.)

Characteristics of River Water

(i) **Dissolved oxygen (DO:** It is the amount of oxygen dissolved in given quantity of water at a particular pressure and temperature.

Significance of DO:

(a) It is vital for the support of fish and other aquatic life in river water.

(b) It determines whether the biological changes are brought about by aerobic or anaerobic micro-organisms.

(c) A minimum level of DO (4mg/lit) must be maintained in rivers so as to support the aquatic life in a healthy condition.

(ii) Biochemical oxygen demand (BOD): It is the amount oxygen required for the biological decomposition of organic matter present in the water.

Significance of BOD:

(a) It is an important indication of the amount of organic matter present in the river water.

(b) Since complete oxidation occurs in the indefinite period, the reaction period is taken as 5days at 20°C. It is written as BOD_5

(c) The rate of oxidation and demand depends on the amount and type of organic matter present in river water.

(iii) Chemical oxygen demand (COD): It is the amount oxygen required for the chemical oxidation of organic matter using oxidizing agents like potassium dichromate and potassium permanganate.

Significance of COD:

(a) It is carried out to determine the pollution strength of river water.

(b) It is a rapid process and takes only 3 hours.

3.4. Soil Pollution

It is defined as "the contamination of soil by human and natural activities which may cause harmful effects on living beings".

Sources

Industrial Wastes

Sources and effects: pulp and paper mills, chemical industries, oil refineries , sugar factories etc., These pollutants affect and alter the chemical and biological properties of soil. As a result, hazardous chemicals can enter into human food chain from the soil; disturb the biochemical process and finally lead to serious effects.

Urban Wastes

Sources and effects: Plastics, Glasses, metallic cans, fibers, papers, rubbers, street sweepings, and other discarded manufactured products. These are also dangerous.

Agricultural Practices

Sources and effects: Huge quantities of fertilizers, pesticides, herbicides , weedicides are added to increase the crop yield. Apart from these farm wastes, manure, slurry, are reported to cause soil pollution.

Radioactive Pollutants

Sources and effects: These are resulting from explosions of nuclear dust and radioactive wastes penetrate the soil and accumulate thereby creating land pollution.

Biological Agents

Sources and effects: Soil gets large quantities of human, animal and birds excreta which constitute the major source of land pollution by biological agents.

Effects

- Ultimately affects the human health by entering into the food chain.
- Affects the soil fertility.
- Decomposition of many types of pathogenic bacteria causes various types of diseases.
- Impact of radioactive elements will cause abnormalities.
- Agricultural purpose results in eutrophication.

Control Measures of Soil Pollution

The soil pollution can be controlled by,

- **Control of soil erosion**: Soil erosion can be controlled by a variety of forestry and farm practices. Reducing deforestation and substituting chemical manures by animal wastes would also help to arrest soil erosion in the long term. Maintaining soil productivity is vital and essential for sustainable agriculture.
- **Proper dumping of unwanted materials**: The excess of waste products by man and animals cause chronic disposal problem. Open dumping is most commonly practiced method. Recently controlled tripping is followed for solid waste disposal. The surface so obtained then can be used for housing or sports field.
- **Production of natural fertilizers**: Excessive use of chemical fertilizers and insecticides should be avoided. Bio-pesticides should be used in place of toxic chemical pesticides.
- **Proper Hygienic condition**: People should be trained regarding the sanitary habits.
- **Public awareness**: Informal and formal public awareness programs should be imparted to educate people on health hazards by environmental pollution.
- **Recycling and Reuse of wastes**: To minimize soil pollution, the wastes such as paper, plastics, metals, glasses, organics, petroleum products and industrial effluents etc., should be recycled and reused.

- **Ban on Toxic chemicals**: ban should be imposed on chemicals and pesticides like DDT, BHC etc. which are fatal to plants and animals. Nuclear explosions and the improper disposal of radioactive waste should be banned.

3.5.   Solid Waste Management

Management of solid waste is very important in order to minimize the adverse effects of solid wastes.

Types of Solid Wastes

Urban (or) Municipal Wastes

- Domestic wastes – It consists of a variety of materials thrown out from the homes. Eg. Food waste, Cloth, Waste paper etc
- Commercial wastes – It includes the waste coming out the shops, markets, hotels, offices, institutions etc. Eg. Packing material, cans, bottles, polyethene etc.
- Construction Wastes – It includes the wastes of construction materials. Eg. Wood, concrete debris etc.
- Biomedical wastes – It includes mostly the waste organic materials. Eg. Anatomical wastes , infectious wastes etc.,

Industrial Wastes

- Nuclear power plants – It generates radioactive wastes.
- Thermal power plants –It produces fly ash in large quantities.
- Chemical industries – It produces large quantities of hazardous and toxic materials.

Hazardous Wastes

- Toxic waste – These are poisonous even in very small or trace amounts.
- Reactive waste- These waste react vigorously with air, water, heat and generate toxic gases.eg. Gunpowder, nitroglycerin.
- Corrosive waste- These wastes destroy materials and living tissues by chemical reaction. Eg. acids and bases.
- Radioactive waste- These are from nuclear power plants and persist in the environment for thousands of years.
- Infectious waste- It causes infections to others. Eg. Used bandage, human tissue from surgery, needles etc.
- Heavy metals- lead, mercury and arsenic are hazardous substances.

- Acute toxic- These wastes have an immediate effect on humans or animals causing death.
- Chronic toxicity – These wastes have long-term effect slowly causing irreparable harm to the exposed persons. It is much more difficult to determine.

Steps Involved in Solid Waste Management

Two important steps of solid waste management are

(i)   Reduce, Reuse and Recycle, before destruction and safe storage of wastes.

(ii) Discarding wastes.

Reduce, Reuse and Recycle (3R)

(a) Reduce the usage of raw materials – If the usages of raw materials are reduced, the generation of waste also gets reduced.

(b) Reuse of waste materials – (i) the refillable containers, which are discarded after use, can be reused. (ii) Rubber rings can be made from the discarded cycle tubes, which reduce, the waste generation during manufacturing of rubber hands.

(c) Recycling of materials – recycling is the reprocessing of the discarded materials into new useful products.

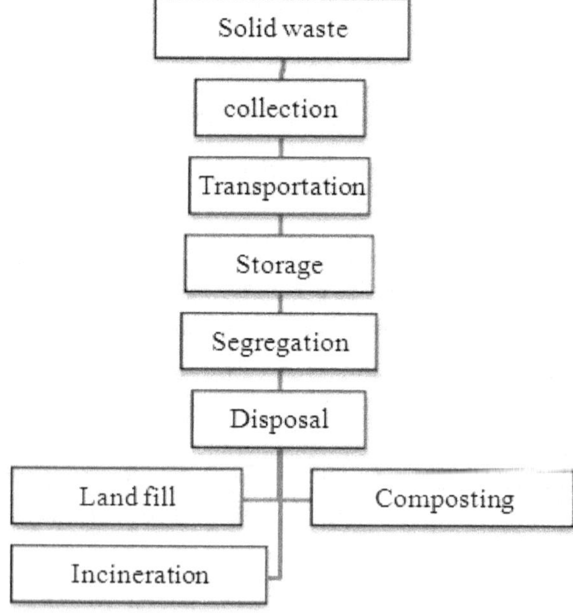

Disposal Methods

Land Fills

Solid wastes are placed in sanitary landfill system in alternate layers of 80 cm thick refuse, covered with selected earth fill of 20cm thickness. After two or three years, solid waste volume shrinks by 25-30% and the land is used for parks, roads, and small buildings. In modern landfills, the bottom is covered with an impermeable liner usually of clay, thick plastic, and sand which prevents the percolation of leachates. Methane produced by anaerobic decomposition is collected and burnt to produce electricity or heat.

Advantage

- It is simple and economical.
- Segregation not required.
- Land filled areas can be reclaimed and used for other purposes.
- Converts low-lying, marshy wasteland into useful areas.
- Natural resources are returned to the soil and recycled.

Disadvantage

- A large area is required.
- Since land is available away from town, transportation cost is heavy.
- Bad odours, if landfills are not properly managed.
- Causes fire hazard due to the formation of methane in wet weather.

Incineration

It is a hygienic way of disposing of the solid waste. It is a thermal process and is very effective for detoxification of all combustible pathogens. During incineration high levels of dioxins, furans, lead, cadmium may be emitted with the fly ash. It is better to remove batteries containing heavy metals and plastic containing chlorine before burning the material. Prior removal of plastics will reduce emissions of toxic gases.

In this method, the municipal solid wastes are burnt in a furnace called incinerator. The combustible substances such as rubbish, garbage, dead organisms and the non-combustible matter such as glass, porcelain, metals are separated before feeding into the incinerators. The non-combustible materials can be left out for recycling and reuse. The left out ashes and clinkers from the incinerators may be accounted for only about 10 to 20% which needs further disposal either by sanatory landfill or by some other means.

The heat produced in the incinerator during the burning of refuse is used in the form of steam power for generation of electricity throughout turbines.

The municipal solid waste is generally wet but has a very high calorific value so it has to be dried up before burning.

The waste is dried in preheater from where it is taken into large incinerating furnace called destructors which can incinerate about 100 to 150 tonnes per hour. The temperature normally maintained in a combustion chamber is about 700°c and may be increased to about 1000°C when electricity is to be generated.

Advantage

- The residue is only 20-25% of original weight; the clinker can be used after treatment.
- It requires very little space.
- The cost of transportation is not high as incinerators located within city limits.
- Safest from a hygienic point of view.
- An incinerator plant of 300 tonnes per day capacity can generate 3MW power.

Disadvantage

- Its capital and operation cost is high.
- Needs skilled personnel.
- Formation of smoke, dust and ashes need further disposal, due to which air pollution may be caused.

Composting

It is another popular method practiced in many cities in our country. In this method, bulk organic waste is converted into fertilizing manure by biological action which improves the soil condition and fertility. The separated compostable waste is dumped in underground earthen trenches in layers of 1.5m and is finally covered with the earth of about 20cm and left over for decomposition. Sometimes certain micro-organisms such as actinomycetes are introduced for active decomposition.

Within 2 to 3days biological action starts, the organic matters are being destroyed by actinomycetes and lot of heat is liberated increasing the temperature of the compost by about 75°C and finally the refuse is converted to powdery brown coloured odourless mass known as humus and has a fertilizing value which can be used for agricultural field. The compost contains a lot of nitrogen essential for plant growth apart from phosphates and other minerals.

Advantages

- When the manure is added to soil, it increases the water retention and ion-exchange capacity of soil.
- A number of industrial solid wastes can also be treated by this method.
- It can manure be sold thereby reducing the cost of disposing of wastes.
- Recycling occurs.

Disadvantages

- The non-consumables have to be disposed of separately.
- Use of compost has not yet caught up with farmers and hence no assured market.

3.6.   Noise Pollution

It may be defined as "the unwanted, unpleasant or disagreeable sound that causes discomfort for all living beings". The noise measurement is done in decibel units. Different permissible noise levels are recommended by the central pollution control board (CPCB).

Types of Noise

- **Industrial noise**: Highly intense sound is caused by many machines. There exists a long list of sources of noise pollution including different machines of numerous factories, particularly from mechanical saws and the pneumatic drill is unbearable and is a nuisance to public.
- **Transport noise**: The main noise comes from transport. It mainly includes traffic noise, rail traffic, and aircraft noise. The number of road vehicles like motors, scooters, cars, motorcycles, buses, trucks and particularly the diesel engine vehicles have increased enormously in recent years.
- **Neighborhood noise**: This type of noise includes disturbance from household gadgets and community. Common noise makers are musical instruments, TV, VCR, radio, transistors, loudspeakers (during celebrations), construction activities (bore well drilling) etc.

Effects of Noise Pollution

- This affects human health, comfort, and efficiency as it interferes the human communication.
- It causes muscles to contract to lead to nervous breakdown, makes the skin pale, leads to excessive secretion of adrenalin hormone into blood stream which is responsible for high blood pressure, tension

- It affects health efficiency and behavior. It may cause damage to the heart, brain, kidneys, liver and may also produce emotional disturbance.

- In addition to the serious loss of hearing due to excessive noise, impulsive noise also causes psychological and pathological disorders.

- The brain is also adversely affected by loud and sudden noise as that of jet and aeroplane noise etc.

- Ultrasonic sound can affect the digestive, respiratory, cardiovascular systems and semicircular canals of the internal ear. The rate of the heartbeat may also be affected.

- Recently it has been reported that blood is also thickened by excessive noises.

Control and Preventive Measures

- Source control This may include containing the source inside a sound insulating enclosure, acoustic treatment to machine surface, design changes, limiting the operational timings and so on.

- Transmission path intervention- This may include containing the source inside a sound insulating enclosure, construction of a noise barrier or provision of sound absorbing materials along the path.

- Receptor control: This includes protection of the receiver by altering the work schedule or provision of personal protection devices such as earplugs for operating noisy machinery. The measure may include dissipation and deflection methods.

- Oiling – Proper oiling will reduce the noise from the machines.

- Usage of sound absorbing silencers.

- Plant more trees.

- Implementing certain laws.

3.7.   Thermal Pollution

It may be defined as the "addition of excess of undesirable heat to water that makes it harmful to man, animal or aquatic life or otherwise causes's significant departures from the normal activities of aquatic communities in water".

Sources of Thermal Pollution

- **Nuclear power plants**: Nuclear power plants including drainage from hospitals, research institutes, nuclear experiments and explosions, discharged a lot of unutilized heat and traces of toxic radionuclides into nearby water streams. Heated effluents from power plants are discharged at 10°C, higher than the receiving water which affects the aquatic flora and fauna.

- **Coal-fired power plants**: Some thermal power plants utilize coal as fuel while a few plants use nuclear fuel. Coal-fired power plants constitute the major source of thermal pollutants. Their condenser coils are cooled with water from nearby lake or river and discharge the hot water back to the stream increasing the temperature of nearby water to about 15°C.

- **Industrial effluents**: Industries generating electricity like coal powered and nuclear-power plants require huge amounts of cooling water for heat removal. Normally the discharged water from stream-electric power industry using turbo generators will have a higher temperature ranging from 6°C to 9°C than the receiving water.

- **Domestic sewage**: Domestic sewage is commonly discharged into rivers, lakes, canals or streams with or without waste treatment. The municipal sewage normally has a higher temperature than the receiving water. Hence, the anaerobic condition will set up resulting in the release of foul and offensive gases in water. The marine organisms which depend on the dissolved oxygen will die out

- **Hydro – electric power**: Generation of hydroelectric power. Sometimes, results in negative thermal loading in water systems.

Effects of Thermal Pollution

- **Reduction in dissolved oxygen**: Concentration of dissolved oxygen (DO) decreases with increase in temperature of water.

- **Increase in Toxicity**: The rising temperature increases the toxicity of the poison present in water. A 10°C rise in temperature doubles the toxic effect of potassium cyanides while 80°C rise in temperature triples the toxic effects of O-xylene causing massive mortality of fish.

- **Interference with biological activities**: Temperature is considered to be of vital significance to physiology metabolism and biochemical process in controlling respiratory rates, digestion, excretion and overall development of aquatic organisms. The temperature changes totally disrupt the entire ecosystem.

- **Interference with reproduction:** In fishes, several activities like nest building, spawning, hatching migration and reproduction etc., depend on optimum temperature.

- **Direct mortality**: Unutilized heat in water is responsible for direct mortality of aquatic organisms. The increase in temperature exhausts the micro-organisms and shortens this life span. Above a particular temperature, death occurs to fish due to failure in the respiratory system, nervous system process.

- **Food storage for fish**: change in temperature alters the seasonal variation in the type and abundance of lower organisms. The fish may lack the right food at the right time.

Control Measures of Thermal Pollution

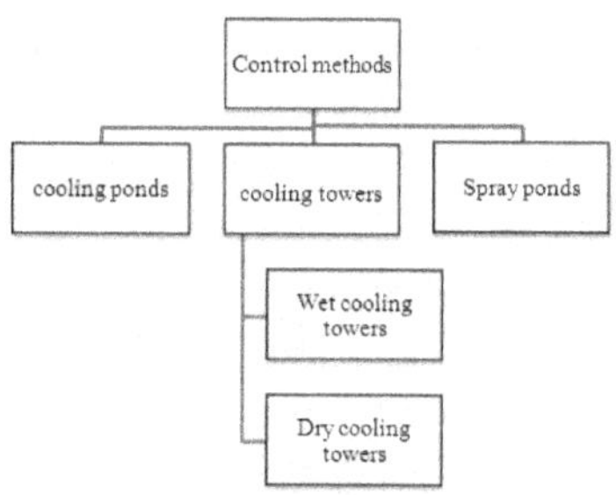

Cooling Ponds

Cooling ponds are the simplest method of cooling thermal discharges. Heated effluents on the surface of the water in cooling ponds maximize dissipation of heat to the atmosphere and minimize the water area and volume. This warm water wedge acts like a cooling pond.

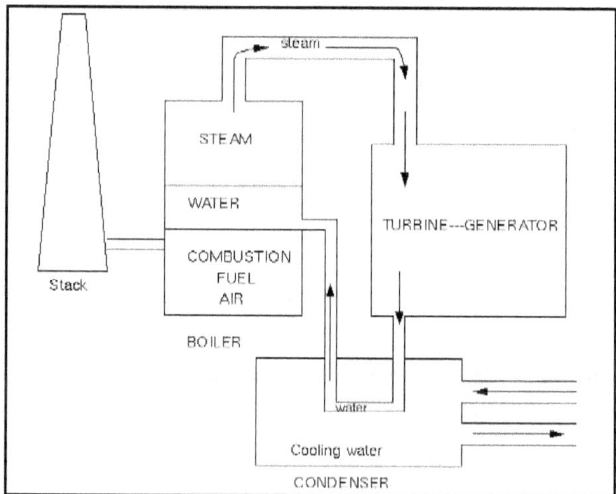

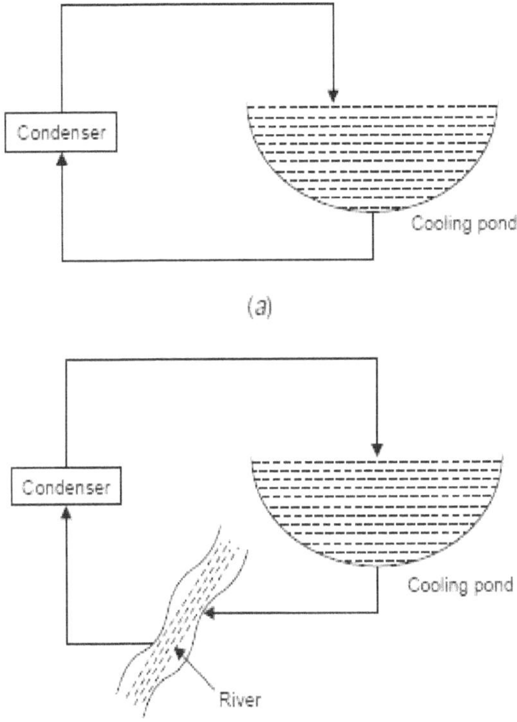

Dissipation of Heat by Cooling Ponds

Cooling Towers

The use of water from water systems for cooling purposes with subsequent return to the waterway after passage through the condenser is termed as a cooling process. To make it more effective, cooling towers are designed to control the temperature of water. Cooling tower transfers some of the heat from hot water to the surrounding atmosphere by the process of evaporation. The cooling tower is generally used to dissipate the recovered waste heat to eliminate the problems of the thermal pollution. The cooling tower is of two types.

a) Wet Cooling Towers

Hot water is sprayed over the baffles. Cool air entering from sides takes away the heat and cools the water which can be recycled or recharged. The drawback concerned with this is the formation of fog which is not good for the environment and causes damage to vegetation.

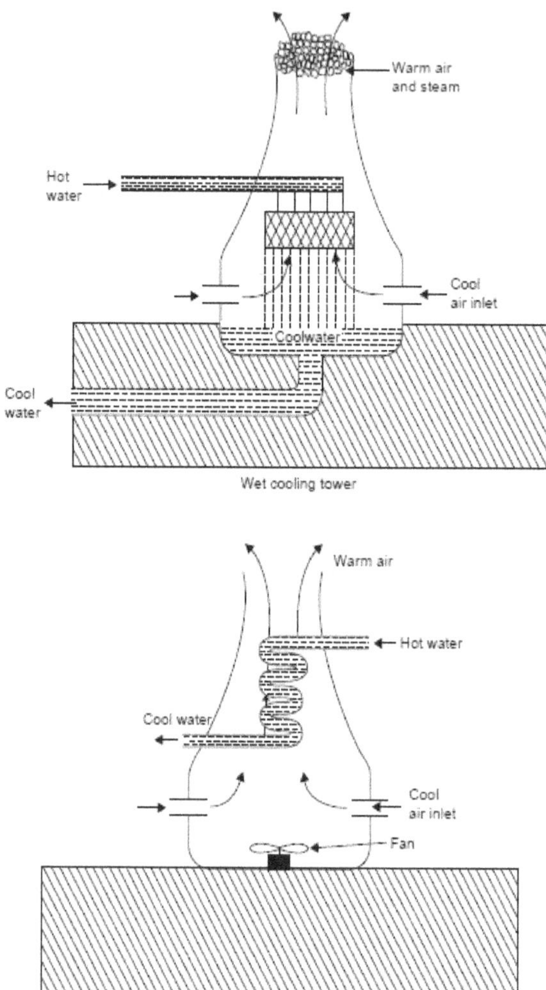

b) *Dry Cooling Towers*

The heated water flows through a system of pipes on which air is passed. Installation and operation cost is many times higher than wet cooling towers. But it is accompanied with less loss of water.

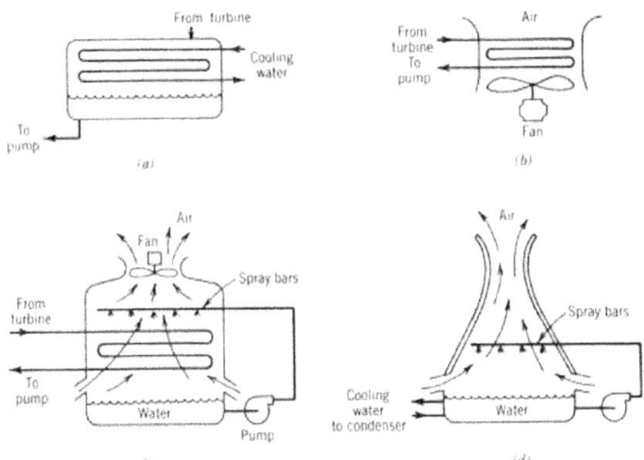

Spray Ponds

The water comes out from the condensers is allowed to pass into the ponds through sprayers. Here the water is sprayed through nozzles where fine droplets. Here from the fine droplets are dissipated to the atmosphere.

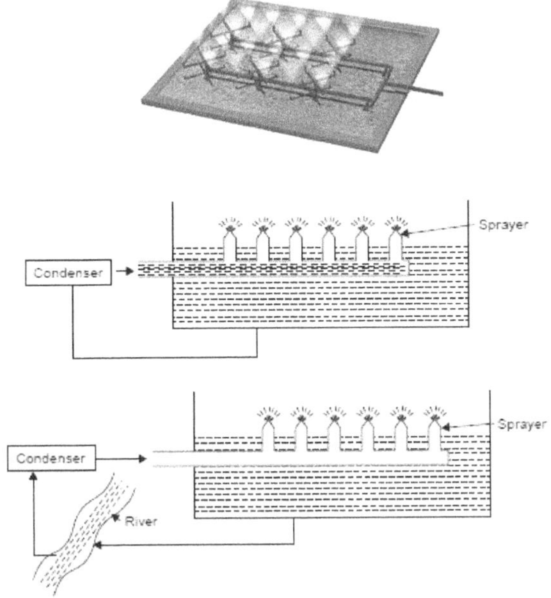

Dissipation of Heat by Spray Ponds

Artificial Lakes

Artificial lakes are man mad bodies of water which offer a possible alternative to once-through cooling. The heated effluents can be discharged into the lake at once end and the water for cooling purposes may be withdrawn from the other end. The heat is eventually dissipated through evaporation.

3.8.   Nuclear Hazards

The radiation hazard in the environment comes from ultraviolet, visible, cosmic rays and microwave radiation which produces a genetic mutation in man. These ionized particles have variable penetration power in which alpha particles have lower penetration power followed by the beta and the gamma radiations being the highest penetration power.

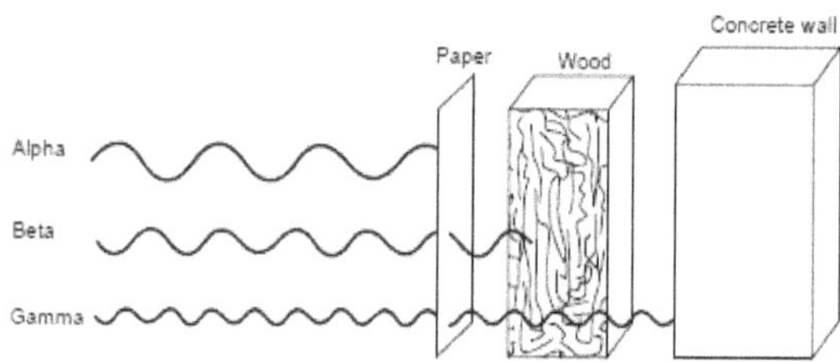

Variable Penetration Power of Ionization Radiations Emitted by Radioisotopes

Sources of Nuclear Hazards

1. Natural Sources – which is in space which emits cosmic rays
2. Man-made sources (Anthropogenic sources) These are nuclear power plants, X-rays , nuclear accidents, nuclear bombs, diagnostic kits etc

Effects of Nuclear Hazards

- Exposure of the brain and central nervous system to high doses of radiation causes delirium, convulsions, and death within hours or days.
- Acute radiation sickness is marked by vomiting, bleeding of gums and in severe cases mouth ulcers results.
- The use of eye is vulnerable to radiation. As its cell die, they become opaque forming cataracts that impaired sight.
- Internal bleeding and blood vessel damage may show up as red spots on the skin.
- Genetic damage is induced by the radiations causing mutations in DNA.
- Somatic damage – It results in burns, miscarriages, eye cataract and bone cancer.

Control Measures

- Nuclear devices should never be exploded in the air.
- In nuclear reactors, closed cycle coolant system with gaseous coolant may be used to prevent extraneous activation products.
- Extreme care should be exercised in the disposal of industrial wastes contaminated with radionuclides.
- Use of high chimneys and ventilations at the working place where radioactive contamination is high seems to be an effective way for dispersing pollutants.
- A minimum number of nuclear installations should be commissioned.
- In nuclear mines, wet drilling may be employed along with the underground drainage.
- Nuclear medicines and radiation therapy should be applied when absolutely necessary with minimum doses.
- Disposal methods are the possible ways to distribute the radio-pollutants. These methods make the pollutant in a confined place to spread over a large space such that pollution can be weakened and its effects can be reduced.

Role of an Individual in Prevention of Pollution

1. Plant more trees and reduce deforestation.
2. Help more in pollution prevention than pollution control.
3. Use water, energy and other resources efficiently and Use renewable energy resources.
4. Purchase recyclable, recycled and environmentally safe products.
5. Remove NO from motor vehicular exhaust and use clean fuel to reduce pollution.
6. Use of eco-friendly products and also use CFC refrigerators.
7. Use rechargeable batteries and do not litter polyethene bags
8. Usage of pesticides under necessary condition.
9. The solid waste generated in one place must be used as raw material for another process.
10. Promote 3R system wherever applicable.

Case Study

Chernobyl Nuclear Disaster

Chernobyl nuclear accident is the worst nuclear disaster in the history of human civilization which occurred at Chernobyl, Ukraine in the erstwhile USSR (now CIS). On 26 April 1986, the accident occurred at the reactor of the Chernobyl power plant designed to produce 1000 MW electrical energy The reactor had been working continuously for 2 years. It was shut down on April 25, 1986, for intermediate repairs. This period coincided with the period when people including the top executives were busy in the preparations for national holiday, The May Day. Due to faulty operations of shutting down the plant, an explosion occurred in the reactor at 01.23 hrs on April 26, 1986. Three seconds later another explosion occurred. The explosion was so severe that the 1000 tonne steel concrete lid of the reactor 4 blew off. The fire started at the reactor due to combustion of graphite rods. The reactor temperature soared to more than 2000°C. Fuel and radioactive debris spewed out in a volcanic cloud of a molten mass of the core and gases. The debris and gases drifted over most of the northern hemisphere. Poland, Denmark, Sweden, and Norway were affected. On the first day of the accident 31 persons died and 239 people were hospitalized. Since the plume was rich in Iodine-131, Cesium-134, and Cesium-137, it was feared that some of the 5,76,000 people exposed to the radiations would suffer from cancer especially thyroid cancer and leukemia. Children were more susceptible as Iodine-131 is ingested mainly through milk and milk products. Since children consume more milk and their thyroid glands are in the growing stage, an increase in thyroid cancer in children from areas near Chernobyl was registered. More than 2000 people died. People suffered from ulcerating skin, loss of hair, nausea and anemia.

Agricultural produce was damaged for years. Intense radiations destroyed several fields, trees, shrubs, plants etc. Flora and fauna were destroyed. Blood abnormalities, hemorrhagic diseases, changes in lungs, eye diseases, cataract, reproductive failure and cancer cases increased. Sweden and Denmark banned the import of contaminated Russian products. The nuclear energy is a cheap, inexhaustible and non-polluting source of energy. However, in the absence of proper care and caution, disasters like Chernobyl can rock the society.

Field study of local polluted site – Urban / Rural / Industrial / Agricultural.

UNIT IV

SOCIAL ISSUES AND THE ENVIRONMENT

4.1. Objectives

Developing and modernizing the technologies without losing our sound traditional values and practices is essential.

Sustainable Development

Meeting the needs of the present, without compromising the ability of future generations, to meet their own needs. This definition was given by the Norwegian Prime minister, G.H.Brunrtland, who was the Director of World Health Organisation.

True Sustainable Development

Optimum use of natural resources with a high degree of reusability, minimum wastage, least generation of toxic by-products and maximum productivity.

Dimensions of Sustainable Development

Multidimensional concept – derived from interactions between society, economy and environment.

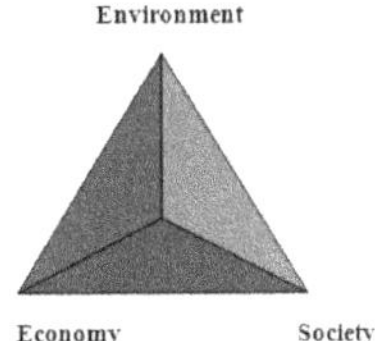

Aspects of Sustainable Development

Inter-generational equity

- This emphasizes that we should minimize any adverse impacts on resources and environment for future generations.
- This can be possible only if we stop over-exploitation of resources, reduce waste discharge and emissions and maintain ecological balance.

Intra-generational equity

- This emphasizes that the development processes should seek to minimize the wealth gaps within and between nations.

- The technological development will support the economic growth of the poor countries and help in narrowing the wealth gap and lead to sustainability.

Measures or Aspects for Sustainable Development

1. Developing appropriate technology.
   - Locally adaptable, eco-friendly, resource efficient and culturally suitable.
   - It mostly involves local resources and local labour.
   - The technology should use less of resources and should produce minimum waste.
2. Reduce, reuse, recycle [3R] approach
   - Reduces waste generation and pollution.
   - The 3R approach advocating minimization of resource use, using them again and again instead of passing it on the waste stream and recycling the materials goes a long way in achieving the goals of sustainability.
   - It reduces pressure on our resources as well as reduces waste generation and pollution.
3. Providing environmental education and awareness
   - This will help to change the thinking and attitude of the people towards our earth and the environment
   - Earth thinking will gradually get incorporated in our thinking and action which will greatly help in transforming our lifestyles to sustainable ones.
4. Consumption of renewable resources
   - Any system can sustain a limited no of organisms on a long term basis which is known as its carrying capacity.
   - If carrying capacity is crossed (by over-exploitation of a resource), environmental degradation starts and continues till it reaches a point of no return.
   - Carrying capacity has 2 basic components:
     1. Supporting capacity – the capacity to regenerate.
     2. Assimilative capacity – the capacity to tolerate different stresses.
5. Conservation of non-renewable resources.
   - Conserved by recycling and reusing.
6. Population control.
   - By controlling population growth, we can make sustainable development.

4.2. Urban Problems Related to Energy

Urbanization

Urbanization is the movement of human population from rural areas to urban areas for the want of better education. Communication, health, employment, etc

1. *Causes for Urbanization*

About 50% of the world population lives in urban areas and there is increasing movement of rural folk to cities in search of employment. The urban growth is so fast that it is becoming difficult to accommodate all the industrial, commercial and residential facilities within a limited municipal boundary.

As a result, there is spreading of the cities on to the suburban or rural areas too, a phenomenon known as urban sprawl.

2. *Energy Demanding Activities*

- Residential and commercial lighting.
- Transportation means including automobiles and public transport for moving from residence to workplace.
- Modern lifestyle using a large no of electrical gadgets in everyday life.
- Industrial plants using a big proportion of energy.
- A large amount of waste generation which has to be disposed oo properly using energy based techniques.
- Control and prevention of air and water pollution which need energy dependent techniques.

3. *Solution for Urban Energy Problem*

- Urban people may use public transport instead of using motorcycles and cars.
- Energy consumption must be minimized in all aspects.
- Production capacity may be increased.
- Use of energy efficient technology.
- Using solar energy and wind energy.
- Imposing strict laws, penalties and energy audit.

Water Conservation

The process of saving water for future utilization is known as water conservation.

Need for Water Conservation

1. Though the resources of water are more, the quality and reliability are not high due to changes in environmental factors.
2. Better lifestyles require more water.
3. the increase in population.
4. Due to deforestation, the annual rainfall is also decreasing.
5. Over exploitation of ground water leads to drought.
6. Agricultural and industrial activities require more fresh water.

Strategies for Water Conservation

1. Reducing evaporation losses.
   - Horizontal barriers of asphalt below the soil surface increase water availability and increase crop yield by 35-40%.This more effective on sandy soil but less effective on loamy sand soils.
   - A copolymer of starch and acrylonitrile called 'super slurper has been reported to absorb water up to 1400 times its weight.
2. Reducing irrigation losses.
   - Use of lined or covered canals to reduce seepage.
   - Irrigation in the early morning or late evening to reduce evaporation losses.
   - Sprinkling irrigation and drip irrigation to conserve water by 30-50%.
   - Growing hybrid crop varieties with fewer water requirements and tolerance to saline water help conserve water.
3. Re-use of water.
   - Treated wastewater can be used for fertile- irrigation.
   - Using grey water from washings, bathtubs etc. for watering gardens, washing cars or paths help in saving fresh water.
4. Preventing of wastage of water.
   - Closing taps when not in use.
   - Repairing any leakage from pipes.
   - Using small capacity flush in toilets.
5. Decreasing run-off losses.
   - Contour cultivation on small furrows and ridges across the slopes trap rainwater and allow more time for infiltration.

- Conservation bench terracing involves the construction of a series of benches for catching the runoff water.
- Water spreading is done by channeling or lagoon leveling. In channeling, the water flow is controlled by a series of diversions with vertical intervals.
- Chemical wetting agents increase the water intake rates when added to normally irrigated soils.
- Surface crop residues, Tillage, mulch, animal residues etc. help in reducing run-off by allowing more water to penetrate into the land.
- Chemical conditioners like gypsum, when applied to sodic soils, improve soil permeability and reduce runoff.
- Water storage structures like farm ponds, dugouts etc. built by individual farmers can be useful measures for conserving water through the reduction of runoff.

6. Avoid discharge of sewage.
7. Increasing block pricing.
   - The consumer has to pay a proportionately higher use of water. This helps in economic use of water by the consumers.

Methods of Water Conservation

1. Rainwater harvesting: A technique of increasing the recharge of ground water by capturing and storing of rainwater for further utilization. This is done by constructing special water – harvesting structures like dug wells, percolation pits, lagoons, check dams etc.

   Objectives of rainwater harvesting.
   1. To reduce runoff loss.
   2. To avoid flooding of roads.
   3. To meet the increasing demands of water.
   4. To raise the water table by recharging ground water.
   5. To supplement groundwater supplies during lean season.
   6. To reducing the ground water contamination.

   Rainwater can be mainly harvested by any one of the following methods:
   1. By storing in tanks or reservoirs above or below ground.
   2. By constructing pits, dug wells, lagoons, trench or check- dams on small rivulets
   3. By recharging the groundwater

Traditional Rainwater Harvesting

In India, it is an old practice in high rainfall areas to collect rainwater from rooftops into storage tanks. In foothills, water flowing from springs are collected by embankment type water storage. In Himalayan foot-hills, people use the hollow bamboos as pipelines to transport the water from natural springs.

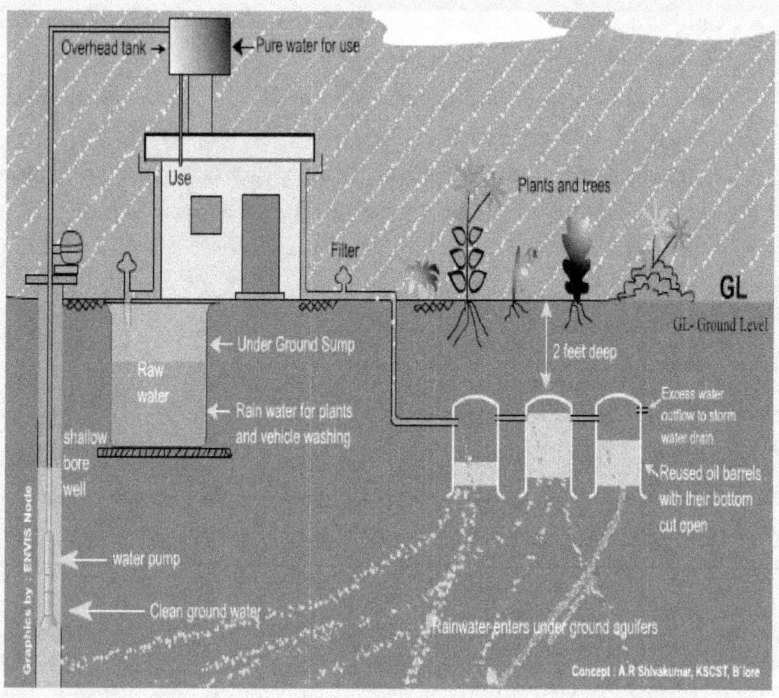

Modern Techniques of Rainwater Harvesting

In arid and semi-arid regions, artificial ground water recharging is done by constructing shallow percolation tanks. Check dams made of any suitable native material are constructed for harvesting runoff from large catchment areas.

Ground water flow can be intercepted by building groundwater dams for storing water underground. As compared to surface dams, groundwater dams have several advantages like minimum evaporation loss, reduced chances of contamination.

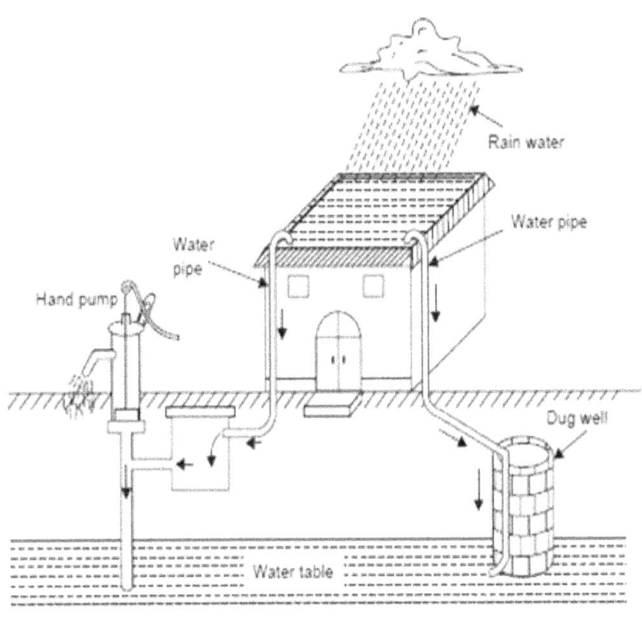

Rain water
Water pipe
Water pipe
Hand pump
Dug well
Water table

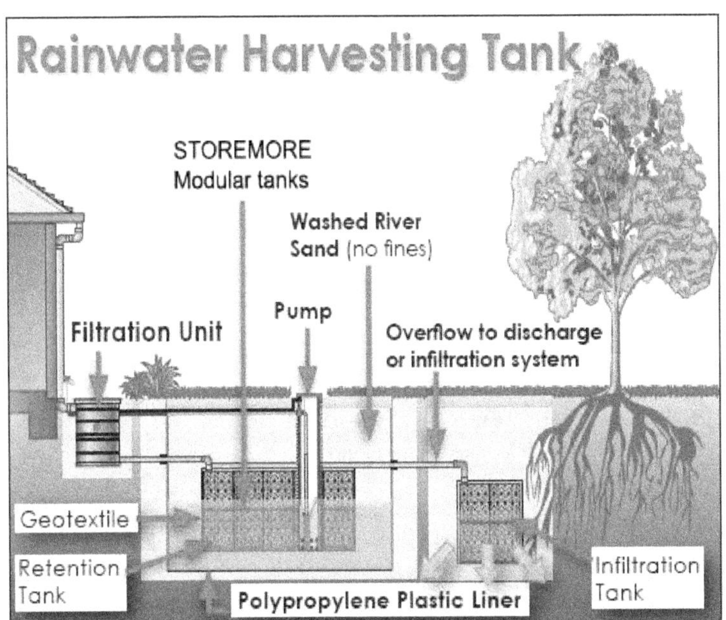

Rainwater Harvesting Tank
STOREMORE
Modular tanks
Washed River
Sand (no fines)
Filtration Unit
Pump
Overflow to discharge
or infiltration system
Geotextile
Retention
Tank
Polypropylene Plastic Liner
Infiltration
Tank

Watershed – The watershed is defined as the land area from which water drains under gravity to a common drainage channel. Thus, the watershed is a delineated area with a well-defined topographic boundary and one water outlet.

Watershed management - The management of rainfall and resultant runoff.

Factors affecting watershed.

- Uncontrolled, Unplanned and unscientific land use.
- Overgrazing, mining, deforestation, construction activities.
- drought climates.

Objectives

1. To minimize risk of floods.
2. For improving the economy.
3. For developmental activities.
4. To generate huge employment opportunities.
5. To promote forestry.
6. To protect soil from erosion.
7. To raise the ground water level.

The concept of watershed management - integrates construction management and budgeting of rainwater through simple but discrete hydrological units.

Watershed Management Techniques

- **Trenches (Pits)** - Trenches were dug at equal intervals to improve groundwater storage.
- **Earthen dam (or) Stone embankment** - Must be constructed in the catchment area.
- **Farm pond** – Can be built to improve water storage capacity of the catchment area.
- **Underground barriers** - Should be built along the nullahs to raise the water table.

Maintenance of Watershed

- **Water Harvesting** - Proper storage of water in watershed is done with provisions that the water can be used in dry seasons in low rainfall areas
- **Afforestation and Agroforestry** – help to prevent soil erosion and retention of moisture in watershed areas.
- **Reducing soil erosion** - Terracing, bunding, contour cropping, strip cropping, etc are used to minimize soil erosion and runoff on the slopes of watersheds
- **Scientific mining and quarrying** – Due to improper mining, the stability of the hills get disturbed resulting in landslides and rapid soil erosion. Planting soil binding plants, contour trenching at an interval of 1m on overburden dump in the mined area are recommended for minimizing the destructive effects of mining in watershed areas.
- **Public participation** – People must be motivated for protecting a freshly planted area and maintaining a water harvesting structure, implemented by the government.
- **Minimizing livestock population.**

Resettlement and Rehabilitation of People

Causes

1. Due to Developmental activities – Construction of dams, mining, roads, airports, urban expansion. These activities cause large-scale displacement of local people from their home and loss of their traditional profession or occupation.

   Eg:

   a) Hirakund Dam: It has displaced more than 20,000 people residing in about 250 villages.

   b) Tehri Dam: It has displaced more than 10,000 residents of Tehri town

2. Due to Disaster - May be of natural or manmade.

   a) Natural disaster - earthquake, floods, droughts, landslides, avalanches, volcanic eruptions etc.,

b) Manmade disaster : industrial accidents, nuclear accidents, dam burst etc.,

3. Due to conservation initiatives – national park, sanctuary, forest reserve, biosphere, etc.,

Resettlement: simple location or displacement of human population. This process does not focus on their future welfare

Rehabilitation: making the system work again by allowing the system to function naturally and includes the lost economic assets, safeguard employment, provide safe land for building, restore social services, repair damaged infrastructures, etc

Rehabilitation Issues

1. Displacement of tribals increases poverty.
2. Break up of families.
3. Communal ownership of property.
4. Vanishing social and cultural activities.
5. Loss of identity among the people.

Case Studies – Sardar Sarovar Dam – about 573 villages consisting 10 lakh people would be homeless and 45,000 hectares of forest and 2,00,000 hectares of cultivated lands would be submerged in Maharastra.

The Theri dam Project – The dam is being constructed across the rivers Bhagirathi and Bhilanganga, close to the Garhwal town of Tehri. The dam would submerge nearly 100 villages, including Tehri, a historical village, 85,600 families will have to be relocated.

Pong Dam- Pond dam was constructed on Beas river in Himachal Pradesh and Punjab. Out of 30,000 families uprooted due to Pong dam, only 16,000 were considered eligible for allotment.

Environmental Ethics

Environmental ethics is the discipline of philosophy that studies the moral relationship of human beings too, and also the value and moral status of, the **environment** and its non-human contents.

Issues

List of Issues in Environmental Ethics

1. Animal Rights.
2. Anthropocentrism.
3. Biocentrism.
4. Conservation.
5. Deep Ecology.
6. Ecofeminism.
7. Moral Standing.
8. Environmental Justice.
9. The Value of Nature.
10. The Land Ethic.
11. Economics and the Environment.
12. Obligations to Future Generations.
13. Sustainable Development and Ethics.
14. Radical Environmental Activism.
15. Intrinsic vs. Instrumental Value in Nature.

Principles

Guidelines relating to human interactions with their environment.

4.3.   Green Chemistry

Definition

Green Chemistry is the utilisation of a set of principles that reduces or eliminates the use or generation of hazardous substances in the design, manufacture, and application of chemical products.

Green Chemistry Is About,

- Waste Minimisation at Source.
- Use of Catalysts in place of Reagents.
- Using Non-Toxic Reagents.
- Use of Renewable Resources.
- Improved Atom Efficiency.
- Use of Solvent Free or Recyclable Environmentally Benign Solvent systems.

The 12 Principles of Green Chemistry

1.   Prevention

It is better to prevent waste than to treat or clean up waste after it has been created.

2.   Atom Economy

Synthetic methods should be designed to maximize the incorporation of all materials used in the process into the final product.

3.   Less Hazardous Chemical Synthesis

Wherever practicable, synthetic methods should be designed to use and generate substances that possess little or no toxicity to people or the environment.

4.   Designing Safer Chemicals

Chemical products should be designed to affect their desired function while minimizing their toxicity.

5.   Safer Solvents and Auxiliaries

The use of auxiliary substances (e.g., solvents or separation agents) should be made unnecessary whenever possible and innocuous when used.

6.   Design for Energy Efficiency

Energy requirements of chemical processes should be recognised for their environmental and economic impacts and should be minimised. If possible, synthetic methods should be conducted at ambient temperature and pressure.

7. **Use of Renewable Feed Stocks**

A raw material or feedstock should be renewable rather than depleting whenever technically and economically practicable.

8. **Reduce Derivatives**

Unnecessary derivatization (use of blocking groups, protection/deprotection, and temporary modification of physical/chemical processes) should be minimised or avoided if possible because such steps require additional reagents and can generate waste.

9. **Catalysis**

Catalytic reagents (as selective as possible) are superior to stoichiometric reagents.

10. **Designs for Degradation**

Chemical products should be designed so that at the end of their function they break down into innocuous degradation products and do not persist in the environment.

11. **Real-time Analyses for Pollution Prevention**

Analytical methodologies need to be further developed to allow for real-time, in-process monitoring and control prior to the formation of hazardous substances.

12. **Inherently Safer Chemistry for Accident Prevention**

Substances and the form of a substance used in a chemical process should be chosen to minimize the potential for chemical accidents, including releases, explosions, and fires.

Wasteland Reclamation

Wasteland: - The land which is not in use – unproductive, unfit for cultivation another economic use. Types of waste land:

1. Uncultivable wasteland – Barren rocky areas, hilly slopes, sandy desserts.
2. Cultivable wasteland- degraded forest lands, gullied lands. Marshlands, saline land etc.,

Causes for Wasteland Formation

1. Soil Erosion, Deforestation, Water logging, Salinity.
2. Excessive use of pesticides.
3. Construction of dams.
4. Over-exploitation of natural resources.
5. Sewage and industrial wastes.
6. Mining.
7. Growing demands for fuel, fodder wood, and food causes degradation and loss of soil productivity.

Objectives of Waste Land Reclamation

1. To improve the physical structure and quality of the soil.
2. To prevent soil erosion.
3. To avoid over – exploitation of natural resources.
4. To conserve the biological resources.

Methods of Waste Land Reclamation

1. Drainage – Excess water is removed by artificial drainage.
2. Leaching – process of removal of salt from the salt affected soil by applying an excess amount of water.
3. Irrigation practices – High-frequency irrigation with a controlled amount of water helps to maintain better water availability in the land.
4. Green manures and biofertilizers - This is to improve the saline soil.
5. Application of Gypsum – Calcium of gypsum replaces sodium from the exchangeable sites. This process converts clay back into calcium clay.
6. Afforestation programmes – The National development Board has decided to bring 5 million acres of waste land annually for firewood and fodder plantation.
7. Social forestry programmes – involve strip plantation on road , canal-sides, degraded forest land, etc

Consumerization of Waste Products

Consumerism Refers to the consumption of resources by the people.

Traditionally favorable rights of sellers- Right to introduce product, price, Incentives. Traditionally buyer rights-Right to buy, right to expect the product to perform as claimed. Important information's to be known by buyers: ingredients, manufacturing dates, expiry etc.

Objectives of Consumerization

1. Improves rights and power of the buyers.
2. Making the manufacturer liable.
3. Reuse and recycle the product.
4. Reclaiming useful parts.
5. Reusable packing materials.
6. Health and happiness.

Sources of Wastes

Glass, papers, garbage's, food waste, automobile waste, dead animals, etc.

E – Waste

Computers, printers, mobile phones, Xerox machines, calculators, etc.

Effects of Wastes

1. Dangerous to human life.
2. Degrade soil.
3. Cadmium in chips, Cathode ray tube, PVC cause cancer and other respiratory problems.
4. Non-biodegradable plastics reduce toxic gases.

Factors Affecting Consumerization and Generation of Wastes

1. People over – population: When there are more people than available food, water, and other resources in an area – causes degradation of limited resources – poverty and under nourishments. Low Developed Countries (LDC) are more prone to these conditions. There is less per capita consumption although the overall consumption is high.

2. Consumption over – population: These conditions occur in more developed countries (MDC). Population size is smaller but the resource consumption is high due to luxurious lifestyle (i.e.) per capita consumption is high. More consumption of resources lead to high waste generation – greater is the degradation of the environment.

   According to Paul Ehrlich and John Holdren model

   Overall environmental impact = no. of people **x** per capita use of resources

   x waste generated per unit of resources

Scheme of Labeling of Environmentally Friendly Products (ECOMARK)

Ecomark or Eco mark is a certification mark issued by the Bureau of Indian Standards (the national standards organization of India) to products conforming to a set of standards aimed at the least impact on the ecosystem.

The Logo for the 'ECOMARK" shall be as notified by the Central Government.

THE CRITERIA FOR ECOMARK: Environmental criteria for each product category will be notified by the Central Government and later on shall be translated into Indian Standards by the Bureau of Indian Standards. The criteria shall be for broad environmental levels and aspects, but will be specific at the product level. Products will be examined in terms of the following main environmental impacts: (a) That they have substantially less potential for pollution than other comparable products in production, usage, and disposal. (b) That they are recycled, recyclable, made from recycled products or biodegradable, where comparable products are not. (c) That they make a significant contribution to saving non-renewable resources, including non-renewable energy sources and natural resources, compared with comparable products. (d) That the product must contribute to a reduction of the adverse primary criteria which has the highest environmental impact associated with the use of the product, and which will be specifically set for each of the product categories.

In determining the primary criteria for a product the following shall be taken (a) Production process including source of raw material; (b) Case of Natural Resources; Likely impact on the environment; (d) Energy conservation in the production of the product; (e) Effect & extent of waste arising from the production process; (f) Disposal of the product and its container; (g) Utilization of "Waste" and recycled materials; (h) Suitability for recycling or packaging (i) Biodegradability; The criteria shall be reviewed from time to time. The draft criteria shall be released for public comments for a period of sixty days.

Emission Standards – ISO 14001 Standard

ISO 14001 sets out the criteria for an Environmental Management System (EMS). It does not state requirements for environmental performance but maps out a framework that a company or organization can follow to set up an effective EMS. It can be used by any organization that wants to improve resource efficiency, reduce waste, and drive down costs. Using ISO 14001 can provide assurance to company management and employees as well as external stakeholders that environmental impact is being measured and improved.ISO 14001 can also be integrated with other management functions and assists companies in meeting their environmental and economic goals.

ISO 14001, as with other ISO 14000 standards, is voluntary (IISD 2010), with its main aim to assist companies in continually improving their environmental performance, while complying with any applicable legislation. Organizations are responsible for setting their own targets and performance measures, with the standard serving to assist them in meeting objectives and goals and in the subsequent monitoring and measurement of these (IISD 2010).

The standard can be applied to a variety of levels in the business, from an organizational level, right down to the product and service level (RMIT University). Rather than focusing on exact measures and goals of environmental performance, the standard highlights what an organization needs to do to meet these goals (IISD 2010).

ISO 14001 is known as a generic management system standard, meaning that it is relevant to any organization seeking to improve and manage resources more effectively. This includes:

- Single-site to large multi-national companies.
- High-risk companies to low-risk service organizations.
- Manufacturing, process, and the service industries, including local governments.
- All industry sectors including public and private sectors.
- Original equipment manufacturers and their suppliers.

All standards are periodically reviewed by ISO to ensure they still meet market requirements. The current version ISO 14001:2004 was last reviewed in 2012. The ISO committee decided a revision was necessary. The new version is expected by the end of 2015. After the revision has been published, certified organizations get a three-year transition period to adapt their environmental management system to the new edition of the standard. The new version of ISO 14001 is going to focus on the improvement of environmental performance rather than to improve the management system itself.

UNIT V

HUMAN POPULATION AND THE ENVIRONMENT

5.1. Human Population

Population: Population is defined as a group of individuals belonging to the same species, which live in a given area at a given time.

Population density: It is expressed as the number of individuals of the population per unit area or per unit volume.

Parameters Affecting Population Size

1. **Birth rate or Natality:** It is the number of live birth per 1000 people in a population in a given year.
2. **Death rate or Mortality:** It is the number of death per 1000 people in a population in a given year.
3. **Immigration:** It denotes the arrival of individuals from the neighbouring population.
4. **Emigration:** It denotes the dispersal of individuals from the original population to the new area.

Population Growth

The "population growth rate" is the rate at which the number of individuals in a population increases in a given time period as a fraction of the initial population.

It is often expressed as a percentage of the number of individuals in the population at the beginning of that period.

Causes of Rapid Population Growth

1. Due to decrease in death rate and increase in birth rate.
2. Availability of antibiotics, immunization, clean water decreases the famine-related death and infant mortality.

Characteristics of Population Growth (or) Factors Affecting Population Growth

- **Exponential growth:** Population growth takes place exponentially $10, 10^2, 10^3, 10^4$ etc that shows a dramatic increase in global population in the past 150 years. When it increases by a fixed percentage it is known as exponential growth.
- **Doubling time:** The time needed for a population to double its size at a constant annual rate is known as doubling time. It is calculated as,

  $Td = 70/r$

  Where Td is the doubling time in years, r is the annual growth rate.
- **Infant mortality rate:** It is the percentage of infants died out of those born in a year. Although this rate has declined in the last 50 years, the pattern differs widely in developed and developing countries.
- **Total fertility rate (TFR):** It is defined as the average number of children that would be born to a woman in her lifetime. The value of TFR varies from 1.9 in developed nations to 4.7 in developing nations.

- **Replacement level:** Two parents bearing two children will be replaced by their offspring. Due to infant mortality, this replacement level is changed. But, due to high infant mortality the replacement level is generally high in developing countries.

- **Male/female ratio:** The ratio of girls and boys should be fairly balanced in a society to flourish. But the ratio has been upset in many countries including China and India. In China, the ratio of girls and boys is 100:140.

- **Demographic transition:** Population growth is generally related to economic development. The death rates and birth rates fall due to improved living conditions. This results in low population growth. This phenomenon is referred to as a demographic transition.

Problems (Environmental Issues) of Population Growth

- Increasing demands food and natural resources.
- Inadequate housing and health services.
- Loss of agricultural lands.
- Unemployment and socio-political unrest.
- Environmental pollution.

Variation of Population among Nations

Different regions of the world find themselves at different stages of demographic transition from high to low mortality and fertility.

Their growth path also differs considerably, resulting in significant shifts in the geographical distribution of world population.

At present, the world's population has crossed 6millions. This existing population is also not evenly distributed, less developed countries have 80% population while the developed countries have only 20%.

Less developed countries (Africa, Asia, and South America) have 80% of the total world population and occupy less than 20% of the total land area.

In the most developed countries like USA, Canada, Australia the population increases at the rate of less than 1% per year. But, in less developed countries like South America, Africa, and Asia, the population increases at the rate greater than 1% per year.

Variation of Population based on Age Structure

It is mainly classified into three classes.

1. Pre-productive population (0-14 years).
2. Reproductive population (15-44 years).
3. Post-reproductive population (above 45 years).

Variation of population is explained based on above three classes,

1. Pyramid shaped – India, Bangladesh, Ethiopia.
2. Bell shaped – France, USA, UK.
3. Urn shaped - Germany, Italy, and Japan.

Pyramid shaped: The Pre productive age group (0-14 years) population is more indicated at the base of the pyramid and post-reproductive age group (above 45 years) is less , indicated at the top of the pyramid. A large number of young age people will soon enter into reproductive age group population (15-44 years), which increases the population growth. But the less number of old age people indicate less loss of population due to death. Example: India, Bangladesh, Ethiopia.

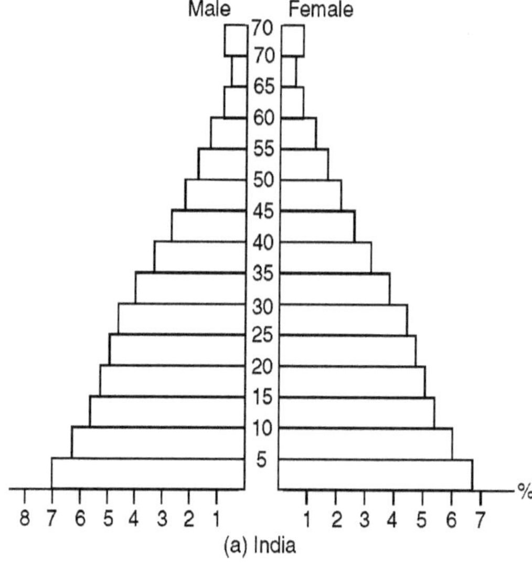

Bell shaped: The Pre productive age group (0-14 years) population and reproductive age group (15-44 years) are more or less equal. The people entering into reproductive age group will not change the population much and such age-pyramids indicate stable population growth. Example: France, USA, UK.

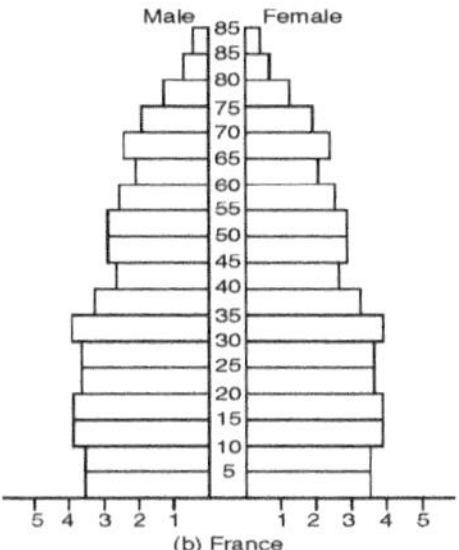

(b) France

Urn shaped: The Pre productive age group (0-14 years) population is smaller than the reproductive age group (15-44 years). In the next 10 years, the number of people in the reproductive age group is less than before resulting in a decrease of population growth.

Example: Germany, Italy, and Japan.

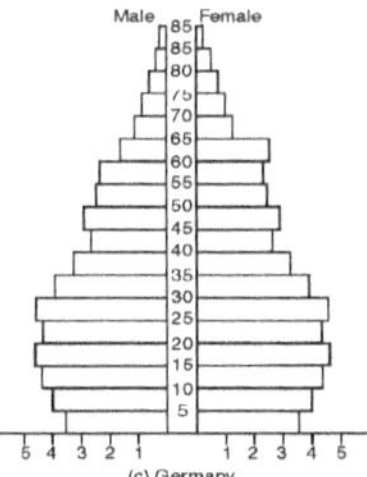

(c) Germany

5.2. Population Explosion (or) Environmental and Social Impacts of Growing Population

Definition

The enormous increase in population due to low death rate and the high birth rate is termed as population explosion.

Causes

1. Modern medical facilities.
2. Life expectancy.
3. Illiteracy.

Effects

1. Poverty leading to infant mortality rate.
2. Leads to environmental degradation.
3. Causes over –exploitation of natural resources.

4. Causes threatening effects to forests, grasses, lands.

5. Develops slum areas due to over population.

6. Lack of basic facilities.

7. Unemployment and low living style.

Remedy: Through birth control programmes the population growth can be minimised.

5.3. Family Welfare Programme

Objectives

1. Slowing down the population explosion.

2. To reduce the overexploitation of natural resources.

Population stabilization ratio: It is obtained by dividing crude birth rate by crude death rate

Family Planning Programme

It predicts the information on birth spacing, birth control and health care systems reducing the number of legal and illegal abortions per year decreasing the death rate.

Objectives (Factors Influencing Family Size)

1. Reduce infant mortality rate.

2. Registration of birth, death, marriage.

3. Encourage late marriages.

4. Improve women's health.

5. Constrain the spread of Sexual diseases.

6. Control of communal diseases.

7. Promote the small family norms.

8. Make school education compulsory to be aware.

Fertility Control Methods

1. Traditional method (Usage of medicines).

2. Modern method (2 methods).

Permanent Method

(It is done by minor surgery).

1. Tubectomy (Female sterilization method).

2. Vasectomy (Male sterilization method).

Temporary Method

1. Condoms.
2. Copper Ts.
3. Injectable drugs.

Family Planning Programme in India

1. India started the programme by 1952.
2. Indian government forced the campaign by 1970's.
3. In 1978, minimum marriage age for men was 18 to 21 years and for women, 15-18 years was declared.
4. Funding for this programme was enhanced.

Environment and Human Health

Environment and human health are the two inseparable entities. If one gets disturbed the other will be affected. Generally, a fit person is a healthy person .But the nutritional, biological, chemical and physiological changes in the human beings brings the diseases.

Factors Affecting Human Health

1. Chemical factors.
2. Biological factors.
3. Nutritional factors.
4. Psychological factors.

5.4.   Hazards and their Health Effect

Biological Hazards

Biological hazards	Health effects
Bacteria , virus , parasites	Malaria , Diarrhea, anemia , cholera , respiratory diseases

Chemical Hazards their Health Effect

Chemical hazards	Health effects
Combustion of Fossil fuels-Particulate matters	Asthma , Bronchitis , lung diseases
Industrial effluents – toxic gases	Causes cancer and death
Pesticides - DDT	Affects food chain
heavy metals – Cd ,Hg , Pb	Contaminated water causing ill effects

Physical Hazards their Health Effect

Physical hazards	Health effects
Radioactive radiations	Affects the cells , causes genetic disorders , causes cancer
UV radiations	Skin cancer
Global warming	Causes famine , mortality

Preventive Measures from Hazards: (Human Health)

1. Wash your hands before eating.
2. Clean the nails.
3. Maintaining the teeth, skin.
4. Drink pure water.
5. Hot food should be consumed.
6. Washing the fruits and vegetables before consumption.
7. Avoid plastic containers and aluminum vessels for food storage.
8. Undergo exercises to be healthy.

Human Rights

Human rights are the fundamental rights which are possessed by all human beings irrespective of their caste, nationality, sex and language.

Universal Declaration of Human Rights (UNDHR)

Some of the human rights that were globally accepted are,

1. Human right to freedom.
2. Human right to property.
3. Human right to freedom of religion.
4. Human right to culture and education.
5. Human right to constitutional remedies.
6. Human right to equality.
7. Human right against exploitation.
8. Human right to food and environment.
9. Human right to good health.

Human Right to Freedom

- Feel free to express the views
- Freedom to form unions or associations.
- Freedom to start any profession.
- Freedom to build houses anywhere legally.

Human Right to Property

Every human being has got the rights to earn property

Human Right to Freedom of Religion

Every human being has got the rights to choose their religion according to their wish

Human Right to Culture and Education

Every human being has got the rights to choose their culture and education. The minority communities like Muslim, Christians, Sikhism can start their own educational institutions.

Human Right to Constitutional Remedies

Every human being has got the rights to obtain the fundamental rights. They have the power to approach the court for their basic rights.

Human Right to Equality

All human beings are equal before laws. There is no any variation between caste , religion , and economic status.

Human Right Against Exploitation

Every human being has got the rights to fight against exploitation. Children should not be employed.

The human right to food and environment-Every human being has got the rights to get sufficient food, clean water, and healthy environment.

Human right to good health.- Every human being has got the rights to get good physical and mental health

Indian Constitution

Indian constitution provides the rights to various sectors.

- **Article 14**: Equality before law.
- **Article 15**: It prohibits discrimination.
- **Article 16**: It provides equal opportunity for all.
- **Article 19**: It provides freedom in all the sectors.
- **Article 20**: It provides protection.
- **Article 22**: Rights of a person in a custody.
- **Article 23**: It prohibits forced labor.

- **Article 24**: It prohibits forced child labor.
- **Article 25**: It provides the freedom to choose the religion.
- **Article 26**: Rights to establish charitable institutions.
- **Article 27**: It prohibits the compulsion of paying tax unnecessarily.
- **Article 28**: It guarantees the secure education.
- **Article 29**: It gives the rights to conserve the minority languages.
- **Article 30**: It guarantees the rights of linguistic minority.
- **Article 32**: It enforces the fundamental laws.

5.5. Value Education

Education is one of the most important tools in bringing about the socio-economic and cultural progress of a country.

The objective of education should not be merely coaching the students to get through the exams with good results and get some good job.

Education does not simply mean acquiring information but using the resources within the limits of ethical value.

Types of Education

1. **Formal education**- All people will read, write and get good jobs and tackle any problems by means of formal education
2. **Value education** – It teaches the differences between right and wrong. To be compassionate, helpful, respectable, tolerant and generous among youths
3. **Value - based environmental education** – It provides knowledge about ecology, biodiversity, resources etc., It creates a sense to preserve the natural resources in a sustainable way.

Objectives (or) Needs of Value Education

1. To improve the integral growth of human being.
2. To create attitudes and improvement towards a sustainable lifestyle.
3. To increase awareness about our national history our cultural heritage, constitutional rights, national integration, community development and the environment.
4. To create and develop awareness about the values and their significance.
5. To know about various living and now- living organisms and their interaction with the environment.

Concept of Value Education

- Why and how we can use fewer resources?
- Why keep our surroundings clean?
- Why have we to use fewer fertilizers and pesticides?
- Why is it important to save water?

Methods of Imparting Value Education

Telling: It is a process of developing values to enable a pupil to have a clear picture of his own situation.

Modelling: It is a method in which a certain individual perceived as ideal values are presented as a model.

Role playing: Bringing out the true feelings of the actor by taking the role of another person without risk.

Problem-solving: A dilemma is presented to the learners asking them what decisions they are going to take.

Studying biographies of great people: This makes the lives with good deeds and thoughts worthy of emulation.

Types of Values

1. **Universal values** – This tells us about the importance of human conditions. They are reflected in life , joy , love , compassion ,tolerance
2. **Cultural values –** This is concerned with right and wrong , good and bad , true and false. It is reflected in language , aesthetics , education ,economics and law
3. **Individual values**-Parents and teachers are the main keys to shaping our individuals. It is reflected in individual goals, relationships, communities., etc
4. **Global values**: If this harmony is disturbed it may lead to an imbalance in ecology and catastrophic results.
5. **Spiritual values:** It is reflected in self-restraint , self-discipline , contentment , reduction of needs etc.,

HIV/AIDS

AIDS-Acquired Immuno Deficiency Syndrome. Acquired means disease is not hereditary but develops after birth from contact with a disease causing agent. Immune deficiency means that the disease is characterized by a weakening of immune system. HIV-Human immunodeficiency virus causes AIDS disease. the virus is passed through infected blood

Transmission of AIDS

- Prostitution.
- Homosexual activity.
- Use of contaminated syringe in blood transfusion and drug addicts.
- Maternal-fetal transmission.

Symptoms

- Persistent fever.
- Fatigue, weakness.
- Diarrhea.
- Weight loss.
- Low number of T cells in blood.
- Swelling lymph nodes, neck.
- Susceptible to infections.

Treatment

- AZT-Azidi thymidine.
- DDI – Dideoxyinosine.

Screening Test

- ELISA-Enzyme Linked Immuno Sorbant Assay.
- Western blot.
- Polymer chain reaction.
- Saliva and urine test.
- Branched DNA test.
- Immunofluorescent antibody assay.

The Major Precautions to Avoid AIDS

- Education.
- Prevention of blood-borne HIV transmission.
- Primary health care.
- Counseling services.
- Drug treatment.

Effects of HIV / AIDS

- A large number of death occurs.
- Loss of labor and production decreases.
- More water is needed for maintaining hygiene in AIDS-affected areas.
- Lack of energy, frequent fever and sweating.

Woman and Child Welfare

Women and child are usually soft who suffer in a number of ways mainly because they are weak, helpless and economically dependent.

Women Welfare

To improve the status of women by providing opportunities for education , employment, and economic independence.

Needs for Women Welfare

1. Women suffer gender discrimination, devaluation at home, at workplace, in matrimony, and in public places.
2. High number of dowry deaths, violence, criminal offences and mental torture to women
3. Male domination violates women.
4. Women are neglected in policy-making and decision-making.

Objectives of Women Welfare

1. To provide education.
2. To impart vocational training.
3. To generate awareness.
4. To improve employment opportunities.
5. To restore dignity, equality and respect.

Various Schemes (or) Various Organizations towards Women Welfare

1. **National network for women and mining (NNWM)** – It is fighting for a gender audit of Indias mining companies.
2. **United nations decade for women:** It propagates several women related issues on international agenda
3. **International convention on the elimination of all forms of discrimination against women (CEDAW)-**It has been created for the promotion of women's human and socio-economic upliftment.

4. **Non - governmental organizations (NGO's) as Mahila manuals:** It creates awareness among women of remote villages to empower them, train them, educate them and helps them to become self-dependent.

5. **Ministry for women and child development:** It aims the upliftment of women by awareness, care, education and health.

Child welfare

Children are the assets of our society of 21 million children born every year, 20 million children are working as child labors in various hazardous industries like match factory, firework industry, pottery divisions etc.,

Reason for child Labors

- **Poverty :** It forces the children to work in unhealthy conditions

- **Want of money:** As parents need money they are in a situation to send their children for a job.

Various Schemes (or) Various Organizations towards Child Welfare

1. **International law** – International standards were formulated to promote and protect the well-being of children of our society

2. **Rights of the child.**
    - Right to survival: Good standards of living , good nutrition, and health.
    - Right to participation: It gives appropriate information to the child.
    - Right to development: It ensures care, support, social security and recreation.
    - Right to protection: Freedom from exploitation, neglecting and inhuman treatment.

3. **World summit on children-**Framed the agenda to be achieved by the children in the beginning of new millennium

4. **Ministry of human resource development (MHRD):** It concentrates mainly on Childs health, education, nutrition, clean and safe drinking water, sanitation and environment.

Centre for science and environment (CSE) – Its report says children consume more water, food and air than adults and hence more susceptible to any environmental contamination.

5.6.   Role of information technology in environment

Information technology plays a vital role in the field of environmental education. It means the collection   and dissemination of information. The internet facilities, information through satellites, world wide web, Geographical information system provides up to date information on environment and weather.

Software for Environmental Education

Many software's are developed as follows.

1.   *Remote Sensing*

It refers getting the information about an object without getting in contact with it. Any field such as astronomy, gravity, acoustics starting from laboratory testing to astronomy can make use of remote sensing.

At present "Remote sensing" is majorly used for detecting the characteristics of the earth.

Remote Sensing System for Resource Management

Nowadays remote sensing is used to extract the information about the natural resources. The depth of the data depends on the experts who analyze the reports.

Example: The remote sensing image of a land can be used to know about the vegetation, soil strength, geographical features.

Applications of Remote Sensing

a) In Agriculture

In India agriculture supports around 70% of the population and contributes 35% of the net product. Remote sensing can provide valuable information for land and water management.

b) In Forestry

Remote sensing provides the information about the forests such as type, density, forest fire, pest, disease induced loss, encroachment, volume of wood, vegetation etc.,

c) In Land Cover

Remote sensing gives the spatial information depending upon the usage in the form of mapping at different scales.

d) Water Resources

Remote sensing provides the data concerned with water management such as surface water body mapping, groundwater targeting, wetland information, flood monitoring, sedimentation, runoff losses, snow cover, irrigation systems etc.,

2. Database

The database is a collection of interrelated data as in the computer where it is arranged in a systematic manner which is easily manageable and quickly solved.

Applications of Database

a) National Management Information System (NMIS)

It provides a database of R & D projects along with the information of research scientists.

b) Environmental information system (ENVIS)

It provides a network of the database in areas like pollution, clean technologies, remote sensing, biodiversity, desertification etc., which functions all about 25 centers.

c) The Ministry of Environment and Forest

They provide database combining various biotic information such as diseases like Malaria, Flour sis, HIV/AIDS.

3. Geographical Information System (GIS)

It is a technique of superimposing various thematic maps using digital data on a large number of interrelated aspects.

Applications of GIS

1. Interpretation of polluted zones.
2. To check environmental problems.
3. Using software's, Maps containing various aspects of water can be designed.

4. Satellite Data

1. Gives correct and reliable information about forest cover.
2. New reserves of oil, minerals and valuable resources can be discovered.
3. Provides information on atmospheric conditions such as monsoon , climate , smog etc.,

5. *World Wide Web (WWW)*

Most of the current data's are available on World Wide Web. Certain online learning centers like www.mhhe.com/environmental science and multimedia digital content manager (DCM) in the form of CD-ROM can be used.

Applications

1. Provides current data on principles, problems, queries and applications of environmental science.
2. It has the recent data's in the form of lectures, powerpoints, animations, quiz, web exercises which are both helpful for teachers and students.

Case Study

Using an environmental calendar of activities:

There are several days of special environmental significant, which can be alebrated in the community and can be used for creating environmental awareness.

Feb 2nd: World Wetland Day is celebrated to create awareness about wetland and their value to mankind.

March 21st: World Forest Day can be used to initiate a public awareness campaign about the extremely rapid disappreance on forests.

March 22nd: World Water Day

April 7th : World Health Day (WHO) come into existence on this day in 1948

April 18th: World Heritage Day; to arrange a visit to a local museum

April 22nd: Earth Day; to draw attention to increasing environmental problems caused by humans

on earth

June 5th : World Environment Day

June 11th: World Population Day

Aug 6th : Hiroshima Day

Sept 16th: World Ozone Day

Sept 28th: Green Consumer Day

October 16th: World Food Day

Glossary

abiotic	:	Non living.
age-structure	:	Percentage of men and women in the young, adult and old stage in a population.
acid rain	:	Toxic gases like SOx and NOx dissolve in rain water to form sulphuric acid and nitric acid and come down as acid rain.
aerobic	:	an organism that needs oxygen to carry on.
anaerobic	:	an organism that lives in the absence of oxygen.
air pollution	:	toxic chemicals, excess heat or noise present in the atmosphere in concentrations that are or may be harmful to humans, other animals or plants.
alley cropping	:	Planting trees and crops alternately (also called agroforestry).
altitude	:	Height above sea-level.
alpha particle	:	Positively charged matter that consists of two protons and two neutrons.
alpha richness	:	Species richness in a small homogeneous area.
ambient air	:	The air surrounding us.
annual	:	Occurring in a year.
aquifer	:	A highly permeable layer of sediment or rock containing water.
arid	:	Dry
atmosphere	:	The mass of air surrounding the earth.
autotroph	:	Organisms that synthesize their own food *e.g.* green plants.
aerosol	:	Minute particles and droplets suspended in the air.
allergens	:	Substances causing allergy.
anthropogenic	:	Human generated; caused by humans.
bioaccumulation	:	Accumulation of non-biodegradable substances in the body.
biodegradable	:	Substances that can be broken down by microbes.
biodiversity	:	Total variability among species of plants, animals and microorganisms.
biogeochemical cycles	:	Cycling of nutrients among living organisms, air, water and soil.
biogeographical area	:	A region with characteristic climatic biological, water and land resources.

biomagnification	:	Increase in concentration of some stable compounds at successive trophic levels in a food chain.
beta diversity or β-richness	:	Variations in species composition across different habitats.
B.O. D.	:	Biological oxygen demand. It is the amount of dissolved oxygen required by microorganisms to break down organic matter present in water.
biomass	:	Organic matter produced by living organisms
biome	:	A broad, regional type of ecosystem with distinct climate, soil conditions, flora and fauna.
biosphere	:	Zone of earth where life is found. It includes air, water and soil.
biosphere reserve	:	World heritage sites identified by IUCN, due to their high biodiversity and unique ecological features, the whole of the ecosystem along with its biodiversity is preserved in these.
biotic	:	Living
bog	:	Water-logged soil usually containing peat.
boreal forests	:	Mixed coniferous and deciduous trees stretching across North America, Europe and Asia.
cancer	:	a disease producing tumors in which cells multiply uncontrollably and invade surrounding tissue.
carcinogen	:	Any agent promoting cancer *e.g.* chemicals, ionizing radiations etc.
carnivore	:	Organism that feeds on other animals.
carrying capacity	:	Maximum population size that a given system can support over a given period of time.
cell	:	The smallest unit of living organisms.
chlorofluorocarbons (CFCs)	:	Chemical compounds with a carbon skeleton and one or more attached chlorine and fluorine atoms; used as refrigerant, solvent, fire retardant and blowing agent.
chemosynthesis	:	Conversion of inorganic substances into organic compounds (by bacteria) in the absence of sunlight.
chlorophyll	:	Green coloured pigment found in green plants.
climate	:	Long term pattern of weather in a particular area.
climax community	:	The ultimate stable community formed during ecological succession, usually a forest.
closed ecosystem	:	Ecosystem having little exchange of nutrients and energy with outside environment.

closed canopy	:	Forests where tree crowns are spread over 20% of the ground.
coliform bacteria	:	Bacteria living in the colon region of human intestine; used as an index of faecal contamination of water.
commensalism	:	A mutualistic relationship in which one organism is benefited while the other is neither benefited nor harmed.
conifers	:	Needle bearing trees producing cones *e.g.* pines.
consumerism	:	Consumption or use of resources.
community	:	Populations of various species living and interacting in a given area.
compost	:	A nutrient rich soil amendment produced by biological degradation of organic material under aerobic conditions
condensation nuclei	:	Tiny particles on which droplets of water vapour can collect.
conservation tillage farming	:	crop farming where soil is least disturbed.
consumer	:	organism who cannot synthesize its own food and get its nutrition by feeding on others.
consumption overpopulation	:	When resource use is at a very high rate resulting in large-scale waste generation and environmental degradation; found in developed nations with less population.
continental shelf	:	Submerged part of a continent.
contour farming	:	Planting across the changing slope rather than in straight lines, to reduce water and soil loss on hills.
contraceptives	:	physical or chemical methods used for family planning.
coral reefs	:	Massive colonies formed by billions of tiny coral animals.
core	:	Inner zone of the earth
crust	:	solid outer zone of the earth
cyanobacteria	:	blue green algae
consumptive use value	:	The direct use values of biodiversity where the product can be harvested and used directly
confined aquifer	:	Aquifer between two relatively impermeable layers of earth.

DDT	:	Dichlorodiphenyl trichloroethane, a pesticide
deciduous	:	Trees that shed their leaves at the end of the growing season.
decomposers	:	Fungi and bacteria that break complex organic matter into simpler molecules and ultimately into inorganic substances.
delta	:	Fan-shaped sediment deposit found at the mouth of a river.
demography	:	Study of human populations.
demographic transition	:	A pattern of falling death rates and birth rates in response to improved living conditions due to industrialization.
desert	:	a biome where evaporation exceeds precipitation.
desertification	:	degradation of once fertile land into a desert like land.
detritivore	:	Organism that consumes organic litter, debris and dung.
dioxins	:	a family of 75 different chlorinated hydrocarbon compounds produced as by-products at high temperature in chemical reactions, usually carcinogenic.
DNA	:	Deoxyribonucleic acid, genetic material.
doubling time	:	Time taken by something (population) to double itself.
drought	:	Condition in which an area does not get enough water due to below normal rainfall.
drip irrigation	:	Use of perforated tubes that give out water dropwise to the soil around each plant.
detritus	:	Dead organic matter.
deuterium	:	Isotope of hydrogen, the nucleus has one proton and one neutron, mass number: 2.
earthquake	:	shaking of ground due to fracturing and displacement of rocks on the earth's crust.
eco-centric	:	a life view advocating moral values and rights both for the human beings and the earth.
ecology	:	Study of interactions of living organisms with their biotic and abiotic environment.
ecological succession	:	The process by which one community is naturally replaced by another one over a period of time.
ecological services	:	Processes or materials provided by ecosystems like pure air, water, nutrients.
ecosystem	:	A biological community and its physical environment exchanging matter and energy.

latitude	:	Distance from the equator.
leaching	:	Process in which various chemicals in upper layers of soil are dissolved and carried to lower layers.
lethal dose	:	The amount of a substance per unit of body weight that kills all the test animals.
life expectancy	:	Average number of years a new born baby is expected to live.
landslides	:	Mass movement of rock or soil down hill.
lithosphere	:	Outer shell of the earth composed of the crust and the rigid outermost part of the mantle.
magma	:	Molten rock below the earth surface.
malnutrition	:	Diet with deficiency of proteins.
mass number	:	Sum of number of neutrons and protons in the nucleus.
matter	:	Anything that has mass *e.g.* nutrients.
mycorrhiza	:	Mutually benificial association between a fungus and roots of higher plants.
monoculture	:	Cultivation of a single crop or tree.
mutagen	:	Chemical or ionizing radiation that cause mutations.
mutation	:	A sudden heritable change.
marsh	:	A wetland without trees.
mulch	:	A protective cover on the ground, may be of dried leaves.
mutualism	:	An association between two organisms so that both of them are benefited, also called symbiosis.
natural gas	:	Underground deposits of gases containing mainly methane and small amounts of propane and butane.
natural hazards	:	Hazards that destroy or damage wild life habitats, damages property and human settlements.
net primary productivity	:	Rate at which plants produce biomass from sunlight.
neutron	:	Elementary particle in the nuclei of all atoms having no electric charge, relative mass $= 1$ (except hydrogen).
niche	:	The functional role and position of a species in an ecosystem *i.e.* what resources it uses, how does it interact with other species etc.
nitrogen fixation	:	Conversion of atmospheric nitrogen gas into ammonia by nitrogen fixing bacteria/cyanobacteria or by electrification.

nuclear fission	:	nuclei of certain isotopes with large mass number are split apart into lighter nuclei when struck by a neutron releasing large amount of energy.
nuclear fusion	:	Two nuclei of isotopes of lighter elements fuse to form a heavier nucleus releasing a large amount of energy.
open sea	:	Part of the ocean beyond the continental shelf.
ore	:	A metal yielding material.
organic farming	:	Farming involving organic fertilizers and natural pest control, no use of inorganic fertilizers and pesticides.
oustees	:	Native people rooted out of their land/home due to developmental activity.
omnivores	:	Organisms that eat both plants and animals.
open canopy	:	A forest where tree crowns cover less than 20% of the ground.
PAN	:	Peroxyacyl nitrate- a group of chemicals causing photochemical smog.
particulate matter	:	Solid particles or liquid droplets suspended in air.
parts per million (ppm)	:	Number of parts of a chemical found in one million parts of a liquid/gas *e.g.* mg/L.
pathogen	:	Organism that causes disease.
peat	:	Semi-decayed organic matter.
perennial species	:	Plants that grow for more than two years.
permafrost	:	A permanently frozen layer of soil in Arctic Tundra.
pH	:	Numeric value that indicates the relative acidity or alkalinity; varies from 0-14 with neutral point at 7; less than 7 is acidic and more than 7 is alkaline.
photochemical smog	:	Mixture of air pollutants (generally coming along with vehicular exhaust) consisting of hydrocarbons and oxides of nitrogen and formed in the presence of sunlight.
photosynthesis	:	Synthesis of food by green plants in the presence of sunlight using carbon dioxide and water.
photovoltaic cell (PV cell)	:	Solar cell that converts solar energy into electricity.
phytoplanktons	:	Small plants like algae, bacteria found floating on the surface of water.
pioneer species	:	The species which colonize the bare soil first of all.
poaching	:	Illegal commercial hunting or fishing.
productive use value	:	The commercially usable values (of biodiversity) where the product is marketed and sold.
point source	:	A single identifiable source that discharges pollutants into the environment.

teratogens	:	Chemicals or other agents that cause abnormalities during embryonic growth and development.
terracing	:	Shaping the land to create level shelves on hill slopes.
thermodynamics	:	A branch that deals with transfer and conversions of energy.
toxins	:	Poisonous chemicals harmful even in small concentrations.
transpiration	:	Loss of water from plant surfaces.
troposphere	:	The layer of air nearest to earth's surface; both temperature and pressure usually decrease in this layer with increasing altitude.
tundra	:	Treeless arctic or alpine biome.
stratosphere	:	Second layer in the atmosphere above troposphere.
urbanization	:	Increasing concentration of population in cities.
unconfined aquifer	:	Groundwater above a layer of earth material with low permeability.
upwelling	:	Movement of nutrient rich bottom water to ocean's surface.
vertebrates	:	Animals with backbones.
volcano	:	Emission of magma from a fissure/vent in earth's surface releasing liquid lava and gases.
water logging	:	Saturation of soil with irrigation water or excessive precipitation so that water table rises close to surface.
water shed	:	The land area from which water drains under gravity to a common drainage channel.
weather	:	Description of physical conditions of the atmosphere.
pollution	:	Environmental condition in which certain substances (including the normal constituents in excess) are present in concentrations that can cause undesirable effects on man and his environment.
wetlands	:	Ecosystems with standing water and having rooted vegetation.
wild life	:	Undomesticated life forms.
X-ray	:	Very short wavelength rays, useful in medical diagnosis. Can cause mutations.
zero population growth (ZPG)	:	When births and immigration in a population just equals deaths and emigration.
zooplankton	:	Small floating animals on surface of water feeding on phytoplanktons.

UNIT VI

GREEN COMPUTING

6.1. Introduction

Green computing or green IT, refers to environmentally sustainable computing or IT. In the article Harnessing Green IT: Principles and Practices, San Murugesan defines the field of green computing as "the study and practice of designing, manufacturing, using, and disposing of computers, servers, and associated subsystems—such as monitors, printers, storage devices, and networking and communications systems—efficiently and effectively with minimal or no impact on the environment." The goals of green computing are similar to green chemistry; reduce the use of hazardous materials, maximize energy efficiency during the product's lifetime, and promote the recyclability or biodegradability of defunct products and factory waste. Research continues into key areas such as making the use of computers as energy-efficient as possible, and designing algorithms and systems for efficiency-related computer technologies.

Origin

In 1992, the U.S. Environmental Protection Agency launched Energy Star, a voluntary labeling program which is designed to promote and recognize energy-efficiency in monitors, climate control equipment, and other technologies. This resulted in the widespread adoption of sleep mode among consumer electronics. The term "green computing" was probably coined shortly after the Energy Star program began; there are several USENET posts dating back to 1992 which use the term in this manner. Concurrently, the Swedish organization TCO Development launched the TCO Certification program to promote low magnetic and electrical emissions from CRT-based computer displays; this program was later expanded to include criteria on energy consumption, ergonomics, and the use of hazardous materials in construction.

Approaches to Green Computing

In the article *Harnessing Green IT: Principles and Practices*, San Murugesan defines the field of green computing as "the study and practice of designing, manufacturing, using, and disposing of computers, servers, and associated subsystems—such as monitors, printers, storage devices, and networking and communications systems—efficiently and effectively with minimal or no impact on the environment."[1] Murugesan lays out four paths along which he believes the environmental affects of computing should be addressed: Green use, green disposal, green design, and green manufacturing.

Modern IT systems rely upon a complicated mix of people, networks and hardware; as such, a green computing initiative must cover all of these areas as well. A solution may also need to address end user satisfaction, management restructuring, regulatory compliance, and return on investment (ROI). There are also considerable fiscal motivations for companies to take control of their own power consumption; "of the power management tools available, one of the most powerful may still be simple, plain, common sense."

Product Longevity

Gartner maintains that the PC manufacturing process accounts for 70 % of the natural resources used in the life cycle of a PC.. Therefore, the biggest contribution to green computing usually is to prolong the equipment's lifetime. Another report from Gartner recommends to "Look for product longevity, including upgradability and modularity." For instance, manufacturing a new PC makes a far bigger ecological footprint than manufacturing a new RAM module to upgrade an existing one, a common upgrade that saves the user having to purchase a new computer.

Algorithmic Efficiency

The efficiency of algorithms has an impact on the amount of computer resources required for any given computing function and there are many efficiency trade-offs in writing programs. As computers have become more numerous and the cost of hardware has declined relative to the cost of energy, the energy efficiency and environmental impact of computing systems and programs has received increased attention. A study by Alex Wissner-Gross, a physicist at Harvard, estimated that the average Google search released 7 grams of carbon dioxide (CO_2). However, Google disputes this figure, arguing instead that a typical search produces only 0.2 grams of CO_2.

Resource Allocation

Algorithms can also be used to route data to data centers where electricity is less expensive. Researchers from MIT, Carnegie Mellon University, and Akamai have tested an energy allocation algorithm that successfully routes traffic to the location with the cheapest energy costs. The researchers project up to a 40 percent savings on energy costs if their proposed algorithm were to be deployed. Strictly speaking, this approach does not actually reduce the amount of energy being used; it only reduces the cost to the company using it. However, a similar strategy could be used to direct traffic to rely on energy that is produced in a more environmentally friendly or efficient way. A similar approach has also been used to cut energy usage by routing traffic away from data centers experiencing warm weather; this allows computers to be shut down to avoid using air conditioning.

Virtualization

Computer virtualization refers to the abstraction of computer resources, such as the process of originated with the IBM mainframe operating systems of the 1960s, but was commercialized for x86-compatible computers only in the 1990s. With virtualization, a system administrator could combine several physical systems into virtual machines on one single, powerful system, thereby unplugging the original hardware and reducing power and cooling consumption. Several commercial companies and open-source projects now offer software packages to enable a transition to virtual computing. Intel Corporation and AMD have also built proprietary virtualization enhancements to the x86 instruction set into each of their CPU product lines, in order to facilitate virtualized computing.

Terminal Servers

Terminal servers have also been used in green computing. When using the system, users at a terminal connect to a central server; all of the actual computing is done on the server, but the end user experiences the operating system on the terminal. These can be combined with thin clients, which use up to 1/8 the amount of energy of a normal workstation, resulting in a decrease of energy costs and consumption. There has been an increase in using terminal services with thin clients to create virtual labs. Examples of terminal server software include Terminal Services for Windows and the Linux Terminal Server Project (LTSP) for the Linux operating system.

Power Management

The Advanced Configuration and Power Interface (ACPI), an open industry standard, allows an operating system to directly control the power-saving aspects of its underlying hardware. This allows a system to automatically turn off components such as monitors and hard drives after set periods of inactivity. In addition, a system may hibernate, where most components (including the CPU and the system RAM) are turned off. ACPI is a successor to an earlier Intel-Microsoft standard called Advanced Power Management, which allows a computer's BIOS to control power management functions.

Some programs allow the user to manually adjust the voltages supplied to the CPU, which reduces both the amount of heat produced and electricity consumed. This process is called undervolting. Some CPUs can automatically undervolt the processor depending on the workload; this technology is called "SpeedStep" on Intel processors, "PowerNow!"/ "Cool'n'Quiet" on AMD chips, LongHaul on VIA CPUs, and LongRun with Transmeta processors.

Operating System Support

The dominant desktop operating system, Microsoft Windows, has included limited PC power management features since Windows 95. These initially provided for stand-by (suspend-to-RAM) and a monitor low power state. Further iterations of Windows added hibernate (suspend-to-disk) and support for the ACPI standard. Windows 2000 was the first NT based operation system to include power management. This required major changes to the underlying operating system architecture and a new hardware driver model. Windows 2000 also introduced Group Policy, a technology which allowed administrators to centrally configure most Windows features. However, power management was not one of those features. This is probably because the power management settings design relied upon a connected set of per-user and per-machine binary registry values, effectively leaving it up to each user to configure their own power management settings.

This approach, which is not compatible with Windows Group Policy, was repeated in Windows XP.

The reasons for this design decision by Microsoft are not known, and it has resulted in heavy criticism Microsoft significantly improved this in Windows Vista by redesigning the power management system to allow basic configuration by Group Policy. The support offered is limited to a single per-computer policy.

The most recent release, Windows 7 retains these limitations but does include refinements for more efficient user of operating system timers, processor power management and display panel brightness. The most significant change in Windows 7 is in the user experience. The prominence of the default High Performance power plan has been reduced with the aim of encouraging users to save power.

There is a significant market in third-party PC power management software offering features beyond those present in the Windows operating system. Most products offer Active Directory integration and per-user/per-machine settings with the more advanced offering multiple power plans, scheduled power plans, anti-insomnia features and enterprise power usage reporting.

Power Supply

Desktop computer power supplies (PSUs) are generally 70–75% efficient, dissipating the remaining energy as heat. An industry initiative called PLUS certifies PSUs that are at least 80% efficient; typically these models are drop-in replacements for older, less efficient PSUs of the same form factor as of July 20, 2007, all new Energy Star 4.0-certified desktop PSUs must be at least 80% efficient.

Storage

Smaller form factor (e.g. 2.5 inch) hard disk drives often consume less power per gigabyte than physically larger drives.

Unlike hard disk drives, solid-state drives store data in flash memory or DRAM. With no moving parts, power consumption may be reduced somewhat for low capacity flash based devices.

In a recent case study, Fusion-io, manufacturers of the world's fastest Solid State Storage devices, managed to reduce the carbon footprint and operating costs of MySpace data centers by 80% while increasing performance speeds beyond that which had been attainable via multiple hard disk drives in Raid 0.In response, MySpace was able to permanently retire several of their servers, including all their heavy-load servers, further reducing their carbon footprint.

As hard drive prices have fallen, storage farms have tended to increase in capacity to make more data available online.

This includes archival and backup data that would formerly have been saved on tape or other offline storage. The increase in online storage has increased power consumption. Reducing the power consumed by large storage arrays, while still providing the benefits of online storage, is a subject of ongoing research.

Video Card

A fast GPU may be the largest power consumer in a computer.

Energy efficient display options include:

- No video card - use a shared terminal, shared thin client, or desktop sharing software if display required.
- Use motherboard video output - typically low 3D performance and low power.
- Select a GPU based on average wattage or performance per watt.

Display

LCD monitors typically use a cold-cathode fluorescent bulb to provide light for the display. Some newer displays use an array of light-emitting diodes (LEDs) in place of the fluorescent bulb, which reduces the amount of electricity used by the display.

Materials Recycling

Recycling computing equipment can keep harmful materials such as lead, mercury, and hexavalent chromium out of landfills, and can also replace equipment that otherwise would need to be manufactured, saving further energy and emissions. Computer systems that have outlived their particular function can be re-purposed, or donated to various charities and non-profit organizations. However, many charities have recently imposed minimum system requirements for donated equipment. Additionally, parts from outdated systems may be salvaged and recycled through certain retail outlet and municipal or private recycling centers. Computing supplies, such as printer cartridges, paper, and batteries may be recycled as well.

A drawback to many of these schemes is that computers gathered through recycling drives are often shipped to developing countries where environmental standards are less strict than in North America and Europe. The Silicon Valley Toxics Coalition estimates that 80% of the post-consumer e-waste collected for recycling is shipped abroad to countries such as China and Pakistan.

The recycling of old computers raises an important privacy issue. The old storage devices still hold private information, such as emails, passwords and credit card numbers, which can be recovered simply by someone using software that is available freely on the Internet. Deletion of a file does not actually remove the file from the hard drive. Before recycling a computer, users should remove the hard drive, or hard drives if there is more than one, and physically destroy it or store it somewhere safe. There are some authorized hardware recycling companies to whom the computer may be given for recycling, and they typically sign a non-disclosure agreement.

Telecommuting

Teleconferencing and telepresence technologies are often implemented in green computing initiatives. The advantages are many; increased worker satisfaction, reduction of greenhouse gas emissions related to travel, and increased profit margins as a result of lower overhead costs for office space, heat, lighting, etc. The savings are significant; the average annual energy consumption for U.S. office buildings is over 23 kilowatt hours per square foot, with heat, air conditioning and lighting accounting for 70% of all energy consumed. Other related initiatives, such as hotelling, reduce the square footage per employee as workers reserve space only when they need it. Many types of jobs, such as sales, consulting, and field service, integrate well with this technique.

Voice over IP (VoIP) reduces the telephony wiring infrastructure by sharing the existing Ethernet copper. VoIP and phone extension mobility also made hot desking more practical.

Greening of IT

Information Technology (IT) is at the heart of every successful modern business. Without it, success is impossible. Yet, the pervasive deployment of IT has had significant, unintended side effects, namely as a significant contributor to the economically unsustainable worldwide dependence on fossil fuels. The awareness of these side effects, though somewhat late in coming, has led some successful companies to turn to a sustainable practice known as "IT greening." IT greening is about using IT more efficiently to achieve reductions in energy consumption, and therefore, considering the acquisition of energy-efficient IT solutions. Within this book, you can find details on the environmental impact of IT, including data centers' consumption of fossil fuel-based electric energy. In addition, we examine many case studies, extracting lessons learned and best practices for implementing green IT.

IT is so pervasive that energy efficiency through the implementation of green IT has moved to center stage for many companies in their pursuit of environmentally helpful practices. This book provides details on the importance of implementing green IT; the significant and growing role of IT and data centers in the world's consumption of electric energy and carbon footprint; and especially the case studies for "lessons learned" and the best-practice approaches for implementing green IT.

As I mentioned in the Preface, green IT is an ideal way for most companies to make a significant step in reducing their carbon footprint for several reasons. First, for competitive reasons, most companies already refresh their computer hardware—laptops, desktops, servers, and storage devices—every three to four years. That refresh cycle provides a recurring opportunity to buy increasingly energy-efficient technology, such as virtual servers, virtual networks, and virtual data storage. Such virtualization can easily reduce IT power consumption for the replaced equipment by up to 50 percent. (For examples, refer to the Environmental Protection Agency's [EPA] "Report to Congress on Server and Data Center Energy Efficiency" or the reports by Jonathan Koomey listed in the Bibliography.) A second compelling reason to move to green IT is that virtualization technology enables you to reduce equipment and system management costs for your data center. Data center green technology is based on a solid business case—even before we consider the savings due to reduced energy costs. A third reason for moving to green IT is that all large companies are moving to such implementation improvements (in IT virtualization, cloud computing, and so on). In addition to information on IT virtualization, this book also includes information on new energy-efficient cooling technologies that support IT, and the impact of electric utility-rate case incentives and government incentives and regulations on promoting IT energy efficiency.

Green IT has many different aspects. In this book, we use the terms **green IT, green computing**, and **green data centers**. Green IT—as used here—is the most comprehensive because it includes all computing, inside and outside the data center. The emphasis of our discussion is on the business aspects of green IT, so the focus is on what to do, rather than the details of how to do it. However, several chapters, especially the case studies, do give details on how to implement green IT, using best practices based on recent experience and lessons learned through dealing with many companies and organizations throughout the world.

Many technology leaders shrug their shoulders at the mention of climate change in conversation, or they pass on conference panels that use the "green" terminology. But in fact, according to exclusive CIO research, they are beginning to think green. Stricter government regulations, rising energy costs and the growing awareness that sustainability is a real business concern are pushing companies to strategize how they will meet future energy demands and calls for carbon emissions data. Green IT is making inroads in the data center; CIOs are also starting to realize that's only the beginning. Fifty-four percent of IT leaders responding to a *CIO magazine* survey about Green IT report that their organizations have environmental sustainability goals for information technology. In other words, they are trying to reduce IT's impact on the planet.

Why Green IT is Better IT

The report on global warming issued in February by the United Nations' Intergovernmental Panel on Climate Change concludes unequivocally that our planet is getting hotter. And, as we approach Earth Day April 22, the evidence that it's all our fault is stronger than ever.

As we go about our busy lives, running our ever more powerful, ever more ubiquitous computers, we are effectively turning up the planet's thermostat. Gartner estimates that carbon dioxide emissions related to the operation of servers and PCs account for 0.75 percent of the annual global total, and that's before factoring in emissions generated by cooling the boxes. Add to that the emissions generated by telecommunications networks, and IT's contribution to the atmosphere's greenhouse gas load is "probably in excess of 2 percent," says Simon Mingay, research vice president with Gartner, adding that that's "a big number for what is essentially a single device." In fact, emissions tied to that device, the computer, are comparable to the level of greenhouse gasses being produced by all the world's airplanes as they crisscross the skies above us.

Now what are CIOs doing about that?

Green Goes Mainstream

Companies today face a panoply of environmental issues, from how much electricity they consume (produced by power plants that run on fossil fuels) to how they deal with toxic wastes. But in most enterprises the CIO has played a minimal role in decisions that affect the environment. For the most part, no one has asked them to do anything more.

But sooner rather than later, someone—your boss, a big customer or a government agency—is going to want to know what you're doing to comply with, support or advance your company's efforts to become more environmentally responsible. This demand will not stem merely from an altruistic desire to behave responsibly (although that's a fine reason); rather, corporate sustainability (as being cognizant of your impact on the planet is now called) has become a cost of doing business. Global regulations that put limits on toxic chemicals and emissions now reach from the manufacturing floor into the data center.

And having an understanding of and an appreciation for one's effect on the environment increasingly is proving to be good for the bottom line. A recent report concluded that there's a direct correlation between good environmental practices and sound overall management. The report's authors—Marc Orlitzky, a lecturer with the Australian Graduate School of Management, and Frank L. Schmidt and Sara L. Rynes, both professors with the University of Iowa—say good corporate citizenship can help companies build the skills and infrastructure they need to cope in turbulent times, as well as improve efficiency.

Daniel Esty, director of the Center of Business and Environment at Yale University, points to a multitude of examples in which employing IT that attends to the planet helps companies cut costs or raise revenue. "One of the key tools that companies have used to develop an eco-advantage is good data collection," says Esty, coauthor, with Andrew Winston, of Green to Gold: How Smart Companies Use Environmental Strategy to Innovate, Create Value and Build Competitive Advantage. "Having a strong set of metrics and indicators allows companies to manage inputs better, reduce waste and achieve higher productivity," Esty maintains.

The Cost-Benefit Analysis

Charity begins at home, and the CIO's responsibilities start with the IT department. According to Gartner, the pervasiveness of computing equipment in most companies makes the IT department a major source of negative environmental impact.

Mitigating that impact will have its cost. All your hardware contains toxic materials that in many places must be recycled. The European Union recently toughened regulations for disposing of old computers; other nations and many U.S. states also have rules for recycling electronics. Meanwhile, new equipment, designed to be energy efficient and comply with regulations for use of toxic substances, may become more expensive if vendors decide to charge a premium for green products.

But the good news is that every company with a data center has a vein of green waiting to be mined. "Any organization that wants to improve its environmental footprint is going to look at power consumption," notes Mingay. "If the company is a professional services, banking or insurance kind of business, IT will probably be the biggest consumer of power by a long way."

Fixing the Data Center

As Senior Editor Stephanie Over by writes in "Clean, Green Machines," data centers consume between 1.5 percent and 3 percent of all the power generated annually in the United States—at the high end, that's equivalent to the electricity needed to power the state of Michigan for one year. But most IT departments don't have a good grasp of their electricity bills because they don't control the facilities their data centers occupy. One company, VistaPrint, where the IT department does manage its data center facilities, discovered that its electric bill was projected to skyrocket if the company, which sells custom printed products online, grew as predicted.

When Cable & Wireless, which hosts one of VistaPrint's data centers, threatened to charge the company extra to cover the cooling costs for its blade servers, then-CIO Wendy Cebula (now the company's COO) asked her IT operations director to improve energy efficiency. The project eventually led to the construction of a second, more energy-conscious data center in Canada, where hydropower provides a renewable, lower-emissions source of electricity. The company now projects savings of 70 percent on its electricity bills. It will also save $450,000 annually by replacing its blade servers with virtual machines that use more server capacity and consume less power. Within a year, VistaPrint will be reducing the emissions from its data centers to a degree equivalent to taking more than 100 cars off the road for a year.

VistaPrint has benefited in part from the increasingly heated competition among chip makers and server vendors to out-green each other. But which server provides the greatest power consumption savings for the buck depends on how you plan to use it. Existing benchmarks that measure application performance or CPU utilization aren't much help because they don't measure power consumption, notes Brent Kirby, a product manager with AMD. Groups including the Standard Performance Evaluation Corp. and the U.S. Environmental Protection Agency are researching a new generation of benchmarks that address server electricity usage, while an effort among computer manufacturers known as The Green Grid aims to identify best practices for data center power management.

Meanwhile, using more efficient cooling systems, or simply sealing holes in your data center's floor, can reduce energy consumption and, ultimately, greenhouse gas emissions. Your company wins by lowering its energy costs, and the planet wins too.

The Green Advantage

IT's contribution to corporate sustainability doesn't stop at the data center door. New and anticipated environmental regulations are prompting companies to reexamine everything from their business processes to their product lines.

As Associate Staff Writer Katherine Walsh writes in "Can IT Make Your Company Green?", companies are beginning to pay more attention not only to what goes into their products but how those products are made. And whether they're tracking systems for monitoring plant emissions, using databases to analyze material use or implementing operational controls, they need IT to do it. These systems can pay dividends beyond keeping companies out of legal trouble, or having their noncompliant products barred from global markets. At Dow Chemical, CIO David Kepler says technology investments that monitor energy use have saved Dow billions.

Other systems are expected to generate top-line revenue growth by helping companies make better products. Both furniture maker Herman Miller and Timberland, which makes outdoor apparel, cater to environmentally conscious customers (and satisfy regulators) by developing products using recycled materials and eliminating toxics from their production processes. IT integrates the databases that track the use of materials with those that manage the production processes in order to optimize the efficiencies the companies reap from these green initiatives.

You have an advantage if your company has already made environmentally sound practices a priority. In Chicago, the law firm Kirkland & Ellis is constructing a green building, with design parameters that include lower power consumption and more efficient air conditioning. (To see the specs for the building, read "Designed for IT," www.cio.com/inprint.)

"There are regulations, there are savings to the firm and there's just being a good citizen," says Kirkland & Ellis CIO Steve Novak. "If there's any way you can reduce consumption, it's good for all."

If your company has not embraced its environmental responsibilities, it's only a matter of time before it will be forced to or suffer the consequences. "Green issues are going to be climbing up the consumer and investor and media agenda," notes Gartner's Mingay. "At some point in time, this increased focus is going to affect the enterprise."

How Green Data Centers Save Your Money

Last summer when Wendy Cebula was shopping for a new vehicle, energy efficiency and lower emissions topped her list of requirements, along with four-wheel drive (her family lives on a hill). Cebula, then CIO at VistaPrint, a $152 million online supplier of custom print services, eventually chose a hybrid model instead of the traditional SUV. Even though she didn't think the incremental savings on gas would make up for the higher sticker price, she says "it was the right thing to do as a human being."

But as a corporate executive, Cebula, now VistaPrint's COO, can't lead with her heart. The right thing to do is whatever enables the business to grow. Those decisions come down to dollars and sense, not what's best for the planet. But every now and then, the two converge.

In late 2005, Cebula noticed her data center's costs were growing. The company, whose operations are almost completely automated, was adding 100,000 customers a month—and growing at nearly 60 percent a year. There was no sign the demand for data necessary to serve those customers would stabilize. When Cebula and Aaron Branham, VistaPrint's vice president of technology and operations, dug into data center operations in detail, they discovered that energy costs were rising significantly.

If there were some way to lower power costs or increase energy efficiency, they could cut expenses. To Cebula, an avid recycler who tries to impart environmental awareness to her kids, doing so would be a double win. She could do "what's good for the environment and what's good for the bottom line," she explains.

Until recently, the environmental impact of the data center was largely ignored. But today, energy experts estimate that data centers gobble up somewhere between 1.5 percent and 3 percent of all electricity generated in the United States. At the top of the range, that's about the amount of electricity it takes to power the entire state of Michigan for a year. Market research company IDC (a sister company to CIO's publisher) estimates that companies spent $26.1 billion to power and cool servers worldwide in 2005. That's more than was spent to power all the commercial buildings in 17 states—from Delaware to Florida and west to Texas, according to the Department of Energy's most recent energy consumption survey.

And that's not all. According to the Uptime Institute, a consortium of companies devoted to maximizing efficiency and uptime in the data center, more than 60 percent of the power used to cool equipment in the data center is completely wasted. In fact, notes a recent study by the group, energy costs have replaced real estate as the primary data center expense. "Data centers that used to cost $10 million now cost $100 million," says Jonathan Koomey, staff scientist at Lawrence Berkeley National Laboratory. "That kind of expenditure gets C-level attention."

It's garnered government interest too. A federal law enacted in December compels the U.S. Environmental Protection Agency to examine power consumption in data centers, evaluate what technology manufacturers are doing to increase energy efficiency and determine what incentives could convince companies to adopt more energy-efficient technology. The European Union is studying the level of carbon emissions from computer equipment. Down the road, local and federal governments in the United States and abroad may end up penalizing organizations that operate inefficient data centers, according to Rakesh Kumar, Gartner research vice president.

The combination of financial, environmental and legislative pressure will force IT organizations to develop greener data centers, says Kumar. By 2011, Gartner predicts, a quarter of new data centers will be designed for maximum energy efficiency and minimum negative environmental impact. But what that means may vary by organization. "There's no generally accepted, standardized way to build a green data center," says Kumar.

At VistaPrint, becoming green has proven easier than Cebula thought. The company bought more energy-efficient servers and improved utilization in its primary data center in Bermuda, steps that have reduced energy usage by 75 percent. As a result, the company expects to save nearly half a million dollars over three years and estimates it will reduce its output of carbon dioxide emissions by several hundred tons in this year alone. That's equivalent to taking more than 100 cars off the road for a year.

VistaPrint also decided to locate a new data center in Canada, where hydroelectric power—a renewable energy source—keeps power costs stable and has potential to lower VistaPrint's electricity bills by another 70 percent. "We were able to reduce our footprint at a time when it was very important to us financially," says Cebula. "And it's much more green."

High-Density Power Surge

Back in 2000, when VistaPrint founder and CEO Robert Keane moved company headquarters from Paris to Waltham, Mass., the fledgling operation was bringing in just over $6 million in annual sales (the company makes business cards and other printed products to order). During the next five years, the company (which is officially based in Bermuda) saw its revenue explode. VistaPrint outgrew its 7,000 square feet of suburban Boston office space and moved a few miles north to a Lexington, Mass., location eight times larger.

During those years, the company focused on automating and optimizing the product design and manufacturing process at its Venlo, Netherlands, plant. The only concern when purchasing equipment—whether for the company's main Bermuda data center (which runs VistaPrint's website and transaction systems and is hosted by Cable & Wireless) or for the one in Lexington (which supports internal systems and IT production)—was that the equipment work. As a result, VistaPrint procured a hodgepodge of servers: back-end systems, front-end systems and databases each ran on different gear, says Cebula. By 2004, VistaPrint got caught up in the blade craze, purchasing machines from IBM.

In the fall of 2005, shortly before Cebula became CIO, VistaPrint's then-COO, Alex Schowtka, hired Branham as director of IT operations. Branham thought blades were a mistake. The problem was the total power pull. "Blade servers look great on paper," Branham says, "but you start piling them into a rack and suddenly you're out of power, you're out of AC."

VistaPrint wasn't the only company that was getting burned by its decision to buy high-density equipment. With quality data center space priced at a premium, companies sought out more compact gear, and vendors happily obliged. "The focus was on getting as much computer power in as small a package as possible," explains Gartner's Kumar. "Energy was not part of the design mentality."

But blade servers need more power than less-dense hardware. A full rack of high-density servers requires 20 to 30 kilowatts of electricity, while traditional data centers are designed to provide 2 to 3 kilowatts per rack. Meanwhile, according to the Uptime Institute, using high-density equipment triples or quadruples facility cooling costs. Energy prices have risen as well.

CIOs who assumed that data center costs would decline as servers got cheaper and more powerful received a rude awakening. In what Kenneth Brill, founder and executive director of the Uptime Institute, calls "the meltdown of Moore's Law," the energy required to power and cool $1,000 worth of server equipment has skyrocketed from 8 watts in 2000 to 109 watts today—eroding some of the benefits from more powerful chips. In the best-case scenario, says Brill, within five years it could take 157 watts to run the same $1,000 worth of hardware; in the worst case, 1,650 watts. In fact, the blade server's selling point—its size—has become irrelevant for some customers. Data center operators find themselves using only two servers in a rack designed for 10; packing any more than two high-density servers in a rack makes them too hard to cool.

That was VistaPrint's quandary. It was the first Cable & Wireless customer in the Bermuda data center to install blades. And the outsourcer was none too thrilled. Cable & Wireless wanted to limit VistaPrint to one blade server per rack. VistaPrint got the vendor to agree to placing two servers per rack, but talk started of an energy surcharge. For years, most outsourcers charged for data center space based on square footage—a metric that did not take into account electricity costs. As a result, says Lawrence Berkeley's Koomey, the outsourcers created a perverse incentive. "If you charge by the square foot, of course the customers want fully packed racks," Koomey says.

A Virtual Solution

For Branham, joining VistaPrint in September 2005 was like "a trip back in time." He had spent the eight previous years at <u>Monster.com</u> and had seen data center operations expand from one server to 1,000. He had spent the past year working on server virtualization. But Branham saw bigger challenges at VistaPrint. "From afar it looks like a print company. But once you get inside and see what's really going on, it's really a technology company," Branham says, noting that everything from sales to manufacturing to shipping is run largely on custom-built software. "At Monster," says Branham, "we weren't anywhere near this data-focused."

Like Monster in earlier days, VistaPrint wasn't making the most efficient use of its data center. "They picked technologies that were right at the time, not right for where they were going." Branham knew that virtualization would reduce power consumption. But Cebula knew that she wouldn't convince anyone to spend more money on servers without a solid business case. So Branham ran the numbers, focusing on power expenditures.

Calculating the total energy consumption of a server isn't straightforward. How much power a server consumes depends on how you use it. "It's not something that our vendors talk about a lot," says Cebula. The vendors were, nevertheless, able to provide data when asked. Using that information, Branham made the case for a greener data center crystal clear.

Branham calculated that if VistaPrint continued to use blades, this solution (conceived of as four racks housing eight IBM blade servers) would eat up approximately 32,000 watts (4,000 per server) and require 9.1 tons of air conditioning to cool. He estimated that the alternative, using eight HP Proliant DL 585 rack-mounted servers and 110 VMware instances, would require 5,500 watts (50 watts per virtual machine) and just 1.6 tons of AC. In addition, virtualization would enable VistaPrint to make better use of its CPU capacity. As is the case in many companies, VistaPrint was wasting additional energy by using only 20 percent of its server capacity. In a 24-hour period, the typical x86 server is used to only 5 percent or 10 percent of its capacity, says Gartner's Kumar, while energy consumption even in idle mode is 60 percent to 80 percent of the energy consumed in use. But the development team was concerned about stability and performance in a virtual environment.

A year earlier, the IT group had attempted virtualization on the blades. Running VMware on the high-density hardware had been a bust. Memory limitations on the blades limited performance and the number of instances that could be run on each server. So when Branham used the "V"-word again, the development group got nervous. Branham won them over with a pilot project that proved the performance case. Swapping in the new servers in the Bermuda data center would mean eating the investment VistaPrint had made in the blades. But Cebula and the executive management team couldn't argue with the ROI. Over three years, the greener solution would save VistaPrint $450,000—more than the cost of the hardware refresh.

Green PC

Green PC is a nickname for a very efficient, environmentally friendly computer. The "green factor" of the machine depends on its power consumption, size and materials from which it was constructed.

Ways to Make a Computer Greener

- Smaller is better - Less material, less waste.
- Multi-core processor - A computer with a multi-core processor utilizes cores as needed, conserving energy.
- Enable power saving modes -Turning on various power saving modes, like automatic sleep or hibernate, reduces power usage when the computer is idle.
- Efficient power supply -If you build a custom machine, look for an 80 Plus certified power supply.

Green Networking

Green networking is a broad term referring to processes used to optimize networking or make it more efficient. This term extends to and covers processes that reduce energy consumption, as well as processes for conserving bandwidth or any other process that will ultimately reduce energy use and, indirectly, cost.

The issue of green networking has many important applications, especially as energy becomes more expensive and people become more conscious of the negative effects of energy consumption on the environment.

6.2. Understanding the Importance of Green IT in Business

Today, the need for businesses to be environmentally aware is more important than ever before. Being environmentally aware simply means knowing the impact your business is having on the environment. By investing in energy-saving technologies and implementing green policies, you can make your business environmentally-friendly. In addition to having a positive impact on the environment, a green approach to IT and computing can help you to reduce your costs and save money.

When it comes to technology, minimizing the amount of space required for servers and decreasing their impact on the environment is the first step in making your business eco-friendly. A responsible thing to do, going green will have a positive impact on your business. So, what is Green IT and how can it help your business? Let's take a look.

A relatively new concept, Green IT or green computing promotes the efficient and responsible use of technology and reduction of electronic waste. To reduce electronic waste, you need to determine the amount of energy used by you and where you can cut back. Moreover, you should ensure that your practices to dispose of business waste are environmentally friendly. To ensure adherence to Green IT, you should encourage eco-friendly IT practices in your company such as proper recycling habits, power management and virtualization. More importantly, you should look to manufacture environmentally friendly products.

The importance of going green can be gauged by the fact that the government recently suggested compliance regulations that would be used to certify data centers as green. Some of the requirements for data centers included using green technologies, recycling and using low-emission building materials. With the importance of going green done and discussed, let's look at some of the benefits of green computing for your business.

Lower Energy Bills

The most obvious benefit of green computing is lowering energy bills. Green information technology practices can help you to lower your overall power usage. Determining the number of computers, printers, phones and other electronic devices in your workplace, how they're used and how much energy they consumer is the first step in the Green IT practice. Using the aforementioned-information, you can do away with non-essential items which will help you to reduce your overall power usage. This will ultimately result in lower energy bills.

Less Waste

Many companies pay money to get rid of the waste in their workplace. Green information technology involves moving to a paperless system. By moving to a paperless system, you can minimize waste significantly in your workplace. This in turn will help you to reduce your company's carbon footprint and save money.

Reduce Office Expense

You can minimize your general office supply expenses by utilizing more green IT in your business. Paper is an expense that most people are aware of but there are few other expenses that you need to consider. By transitioning to a digital platform, you can reduce expenditure on pens, markers, paperclips and more. How much do your business spend on these items each year? I'm guessing it's in the thousands. If you save money on these expenditures, then the best thing that you can do is move to a digital and green IT system. Also, when you move to a digital platform, the storage required for the aforementioned-items is reduced, which makes room for other equipment and needs.

Improve Workplace Productivity

Increased productivity is another benefit of employing eco-friendly technology. For example, compared to sifting through a filing cabinet, finding digital documents on the Cloud is an easier and faster way of locating the documents you want. Mobile technology and messaging systems are positively impacting how people perform their jobs. Infact, some companies have reported a 25% increase in productivity after moving to a digital platform. Depending on the type of services the company renders, this increase in productivity could lead to an increase in revenue.

Trickle Down Effect of Green IT

The majority of the data centers consume thousand times more power than office space of equal size. Additionally, IT equipment accounts for a major proportion of all energy consumed by businesses. So, the trickle-down effect of transitioning to Green IT impacts the cost of business operations on so many levels. Some companies experience a return of investment of a green IT initiative in three to six months. However, there are cases where the impact is almost immediate. Keep in mind that I've explained the trickle- down effect of Green IT without even touching the cost-saving potential of virtualization and virtualization is just one aspect of Green information technology.

Save Money Across the Board

If you want to save money across the board then, moving to green IT can do you a world of good. You can utilize the savings you achieve through green IT in other aspects of your business to generate even more revenue. For instance, you can use the money saved in energy bills to improve your marketing efforts or products features. Or, you can reinvest that money to further improve your business such as investing more in social media or hiring additional staff.

Tax Benefits

Finally, if you play your cards right, you can enjoy tax benefits after making the switch to Green IT. Depending on what technology you implement, your green devices could help you get a tax break. For instance, by installing solar panels at their workplace, many businesses are enjoying tax savings of as much as 30% and this is just one scenario where switching to green IT results in tax savings.

Small businesses can do their bit for the environment, *and* reduce costs, simply by being a little smarter about how they use information technology. Start here with our top 10 green IT tips, compiled with help from consultant Roger Hyde, of Consult IT Green. They are presented in no particular order.

1. Measure Usage

Examine current business practices to see where you're using energy now. How many computers, printers, phones, scanners, servers, faxes and other electronic devices do you own? How are they used? How much energy do they consume?

Your electric utility may offer free energy audits. Some may also offer networked meters that let consumers (and businesses) see how much energy they're consuming as they consume it. Measuring first not only helps you see where you could save, Hyde said, it also lets you see bottom-line benefits when you start implementing green IT practices.

If you use Intuit's popular QuickBooks small business accounting program (or even if you don't), take a look at the company's Green Snapshot service. It estimates your carbon footprint, provides customized recommendations to help you go greener, and generates we're-a-green-business marketing materials.

2. *Appoint a Green Team*

Make green IT a company-wide responsibility. Appoint a green committee of employees from different areas and task them with developing ideas, plans and objectives. They should meet regularly -- during office hours.

The objective here is not just to delegate the responsibility or to come up with more good ideas than you might have thought of on your own -- although both are potential benefits. It's also a way to ensure that employees feel some ownership for whatever green IT initiatives you eventually decide to take, so they'll be more likely to support them.

A couple of things to keep in mind. Be sure to include employees you know are keen about green issues -- but not only those employees. And be prepared to implement the ideas your team produces.

3. *Consolidate, Outsource*

File this under stating the obvious, but the fewer devices you use, the less energy you use -- and the lower your costs.

Replace separate printers, faxes and scanners with multifunction printers (MFPs) -- once your old devices have worn out. Or use an iPhone instead of a laptop, cell phone and office phone, Hyde suggested. Server consolidation, reducing the number of physical servers by running several logical servers on each machine using virtualization software, is another key initiative. Gartner estimates that every physical server eliminated through consolidation saves 7000kWh of electricity annually, or about $700. You can also consider the ultimate form of consolidation: outsourcing. Use shared data centers, hosted phone systems or cloud computing services. Outsourcers and service providers have to pay close attention to energy conservation -- it's one of their biggest costs -- and they spread energy usage and costs across multiple clients.

4. *Turn Off and Unplug*

The easiest way to reduce energy consumed by IT equipment is to switch it off. Turn off computers at night and configure PC power management utilities to shut down monitors and hard drives when they haven't been used for a few minutes. If you want to see how much you save simply by turning off computers at night, check out this information on the U.S. government's Energy Star program website.

Even during the day, turn off peripherals such as printers and fax machines when no one is using them -- which in many offices is most of the time. One radical suggestion: turn off your computer for a few hours every day.

It saves power, but might also give you more time to think about your business and interact with employees and customers.

But simply switching off equipment is not enough on its own. Many devices draw power even when "off."

Use power strips and unplug them at night (after powering down the attached devices.) Nokia, the cell phone maker, estimates that if its approximately one billion people worldwide unplugged their phones' battery chargers when not in use, it would save enough energy to power 100,000 European homes.

5. Buy Green

If you have to buy new equipment (see below), buy green. Study power consumption ratings for competing products and choose accordingly. Consider integrated devices such as MFPs.

The government's Energy Star program tests and rates every product that makes claims to energy efficiency (and can back them up). Some state governments offer rebates for buying Energy Star-compliant products -- mostly to consumers, but in some cases businesses too.

Just switching from a desktop computer and CRT (cathode ray tube) monitor to a laptop can save between 50 and 80 percent on energy consumption, according to the European Commission's Energy Star directorate.

6. Keep IT Equipment Longer

It's a fine balance. On the one hand, modern Energy Star-compliant equipment consumes less energy and costs less to run. On the other, upgrading to a modern product costs money and means disposing of old equipment, which all too often ends up in landfills.

It's almost always cheaper to add more memory or even a faster processor to an old computer, Hyde pointed out. And do you really *need* that new, faster PC? If you must buy new, donate old equipment or at the very least ensure that you dispose of according to EPA (Environmental Protection Agency) guidelines.

Some computer companies, including Dell and Hewlett-Packard, offer asset recycling programs, taking your old computer and disposing of it ethically when you buy a new one from them. Staples and Office Depot also offer e-waste recycling services.

7. *Telecommute*

Green IT has two aspects: reducing the carbon footprint of IT operations, and using IT to reduce carbon footprint elsewhere.

Telecommuting -- sending employees home to work or to a satellite office closer to home -- is one way to use computers and telecommunications to reduce carbon footprint.

While computers and other electronics draw a lot of power, Hyde noted that the environmental damage and energy consumption associated with "firing up the SUV and driving into work," is far greater.

In a recent study titled 'Work shifting Benefits: The Bottom Line,' Telework Research Network (TRN) estimates that letting one employee work half-time away from the office saves a company about $10,000 a year and the employee up to $6,800.

TRN also estimates that the reduction in driving and gasoline consumption resulting from widespread adoption of telecommuting could cut greenhouse gas emissions nationwide by as much a 53 million metric tons.

8. *Virtual Travel, Real Savings*

A plethora of new, greatly improved and often very inexpensive online technologies make it possible to eliminate most business travel -- without jeopardizing the business.

Consider eliminating or reducing in-person meetings, even with customers, by using low-cost audio conferencing solutions from providers such as Fonality and 8x8, coupled with Web conferencing services such as WebEx. To make online meetings more personal, use VoIP-based (voice over Internet protocol) video conferencing, which costs very little, from providers such as Vidyo, ooVoo or Skype (which is free for two-person video conferences).

Tools such as Central Desktop and Zoho, designed more for non-simultaneous collaboration -- file sharing, collaborative document editing, etc. -- can make it feasible for highly distributed work teams to function smoothly without ever traveling to meet.

9. *Go Paperless*

Information technology was supposed to make this happen decades ago. It still hasn't, mainly because people can't seem to kick the paper habit. The first step: commit to going paperless. Then develop plans for managing the wrenching cultural change. Technology and enlightened policies can do much of the rest.

Reduce the number of printers available to employees. Educate them about the environmental and economic impact of printing. Automatically add a note to emails asking recipients not to print them.

Deliver marketing materials, reports, plans, bills, invoices, etc. electronically. If customers want to print them, they can. You save the cost of printing -- and reduce your carbon footprint. Tell others you can't accept paper, that you don't have filing cabinet space, and paper will end up in the garbage.

10. Print Smart

Sometimes you *have* to print, but even when you do there are ways to reduce the cost and environmental impact.

Adjust printer configurations to ensure margins are the minimum width (to get more print per page), and make double-sided printing the default. When replacing printers, make sure the new ones make it easy to print double-sided.

Use "draft" mode when printing documents internally -- it uses less ink. Re-use paper for internal documents rather than throwing single-sided documents out after use. Finally, consider using Eco-font, a software-based sustainable printing system that uses up to 25 percent less ink. Green IT for small businesses: it's not only easy to do, it's a no-brainer. You save the world and save your company money at the same time.

6.3.   Green IoT

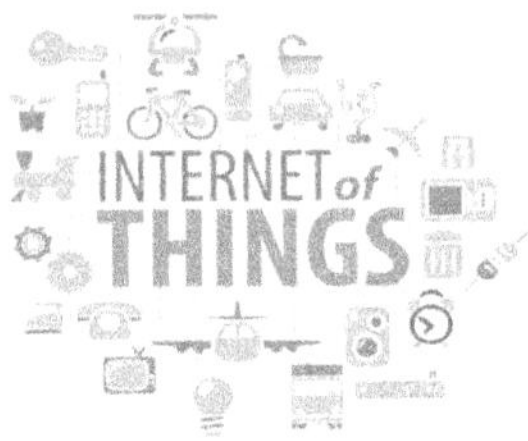

In recent times, the world has been plagued with various environmental concerns. From increased emissions and global warming to the gradual depletion of energy sources, we really are at a critical juncture where a change for the better needs to be made. The Internet of Things is one thing that can contribute majorly to this change. To begin with, it was only computer and individual users who were connected with the internet. Now, the realm of the internet is replete with billions of different devices, many of which are unique and innovative in terms of features and functionality. Although we are at a very early formative stage of the Internet of Things, it is clear that a worldwide network of different machines connected to one network can bring excellent environmental benefits to people, businesses and the world at large.

The Basics of the Internet of Things

To begin with, computers where the only devices that had the capability to send and receive data over a particular network. The Internet of Things is a concept which promotes the idea of devices which are not essentially computers can also be given the same abilities. This concept has been teased in the information technology industry from many years but it is only recently that the development of innovative hardware like sensors, wireless computing technologies, and capable cloud computing platforms that the concept has really started to take hold, creating a mesh of connected smart objects with numerous uses.

Examples of this technology include automobile. It can record and upload performance information to manufacturers servers, sensors attached to water supply pipes that calculate water flow, spot potential problems and help in conservation and electronic home appliances that are directly connected to their producing companies for better management and more pervasive control. Similar capabilities have already become popular in devices like smartphones and tablets, which also offer users with different sensing capabilities that can accurately gauge different types of movement, location etc.

How IoT and Environment Merges Seamlessly

It is true that this connection and interconnection that the Internet of Things has to offer opens the doors to the collection and storage of important data. This data can be used by people, businesses, cities and towns and even entire countries to ensure smoother, more efficient management of its diverse natural resources. One of the most important and environmentally significant aspects of this capability is the monitoring and control of the use of energy. Many companies currently use networked sensors to look at potential problems with their equipment and supply lines and optimize their use of energy to ensure a safer, greener approach.

As more items in devices start getting connected to a particular facility, the managers can get hold of more and more relevant, important data and exercise are mind-boggling level of control over these facilities. The ability to exercise remote control on any or a group of the connected devices in such a network also gives manufacturers the power to conduct excellent preventive maintenance operations. Information from the sensors can be processed and used to ensure that machines deliver optimal performance, and to alert the relevant people when they stopped doing that so that the essential repairs and maintenance can be undertaken and energy-efficient, minimum waste usage can be set into motion.

In cities, towns and countries as well, The Internet of Things can provide a marked to efficiency and improve the general quality of life in many different ways. From monitoring weather conditions to picking out locations where there is the maximum amount of pollution, the use of this technology can come in handy when it comes to taking better care of the environment in both local and wider settings and making a difference to the worldwide environment in a constructive manner. Local governments can monitor effluent disposal, emissions and many more important parameters which can have a lasting impact on the immediate environment, and can take the appropriate measures when they find any of the regulations being flouted.

The Involvement of Big Data

Big Data and the Internet of Things are in essence partners in crime in the way that they facilitate the power and scope of each other. With the numerous concepts and technologies that the Internet of Things brings to the table, there is no wonder that the collection and analysis of large volumes of data go hand-in-hand with the involved processes. Organizations who are currently actively using network devices to monitor their processes, their raw materials and their supply routinely work with large volumes of data which can accurately tell them the location of potential problems, which can help in timely resolution. Similarly, the marriage of the two has been profusely in use in industries creating renewable energy where data from weather forecasts is combined from sensor data to accurately forecast the possible output of windmill farms. The potent combination helps operators to accurately adjust the pitch of the turbine blades in individual windmills to best harvest the power from the wind. Monitoring sensor data with data analytics has already found numerous uses in many different areas covering a whole range of applications from home conveniences to intricate medical procedures.

The Internet of Things and Environmental Sustainability

Having a lasting impact on the physical environment in a way that improves the quality of life of people and brings long-term good effects to overall weather conditions is one of the ways that The Internet of Things can transform the lives of people going into the next decade. Currently, there are cities where smart parking garages fitted with connected sensors are being used to control emission and bring down the level of air pollutants. Similarly, certain cities are in the process of installing sensors in streetlights to facilitate motion controlled operation, which can bring down power bills across the entire city. There are projects where municipalities are installing sensor fitted trash cans which send the message to garbage collectors once the cans are full and ready for clearance.

Using the Internet of Things in a constructive manner with long-term positive impacts on the environment include the use of this technology to measure emission levels from factories and keeping them under statutory control, aiding in agriculture in many different ways and even detecting forest fires in alerting the relevant authorities. Factories and manufacturing plants can also use carefully installed sensors to analyze heat, pressure and other important working variables so as to make the concrete strategy of inserting the right valves and prevent any leakage of energy, thereby contributing to the green movement.

Due to the extremely large and complex array of devices and hardware that is intrinsically connected with the Internet of things, the exact scope and capabilities of this paradigms shift in technology have not yet all been resolved or even thought of. It is a work in progress, and can be treated the same way as things used to be when the Internet first came into being. Even at its nascent stages, the amazing lure of this technology is already visibly front and center in many different sectors of retail, health and medical care and green initiatives such as electricity control, water control, emission control etc.

The energy efficient procedures (hardware or software adopted by IoT either to facilitate reducing the greenhouse effect of existing applications and services or to reduce the impact of greenhouse effect of IoT itself. In the earlier case, the use of IoT will help reduce the greenhouse effect, whereas in the later case further optimization of IoT greenhouse footprint will be taken care. The entire life cycle of green IoT should focus on green design, green production, green utilization and finally green disposal/recycling to have no or very small impact on the environment.

1. **Green RFID** by reducing the size of RFID tags to decrease the amount of non-degradable material used, as the tags are hardly reusable. And also by energy efficient algorithms for tag estimation.
2. **Green WSN**
    a) Triggered powering up of sensor nodes when it is needed, which will save up energy consumption.
    b) Energy depletion by generating power for sensor nodes from environment such as sun, kinetic energy, temperature differentials etc.
    c) Radio optimization techniques.
    d) Minifying the size of data that is transmitted over the communication interface by aggregating, adaptive sampling, compression.
    e) Routing the data in an energy efficient routing technique.

3. **Green CC** by adopting energy efficient hardware and software components, using power saving virtual machine techniques, following energy efficient resource allocation mechanisms etc.

4. **Green M2M** by intelligently adjusting the transmission power, design efficient communication protocols, activity scheduling etc.

5. **Green DC** by sleep scheduling, minimizing the length of data paths, wireless data paths, designing energy aware routing algorithms etc.

Applications of Green IoT

The green IoT which makes the smart devices to communicate to real world efficiently thus focuses on saving of energy and pollution.

The numerous applications of G-IoT are as follows:

Smart Home: A G-IoT enables home equipped with lighting, heating, and electronic devices that can be controlled remotely by smartphone or computer. It can be equipped with Waste removal, Ultrasonic showers, Beds that make/change sheets themselves, Lighting creates artificial sunrise, Computer suggests clothing based on your taste, weather, Windows and walls will allow adjustable amounts of sunlight, warmth or cold in, Electronic soundproof rooms and windows. Soundproof energy fields that you can walk through, Hidden computers, sensors, microphones and electronics throughout the house. Central computer accepts voice commands, distinguishes between occupants for personalized responses and actions, Television, computer and phone merge into one device etc.

Industrial Automation: Industries have been automated with machines that allow for fully automated tasks without or with little manual intervention. An internet based industry automation system that allows a single industry operator to control industry appliances. Smart **Healthcare:** IoT is revolutionizing Healthcare industry by bringing up new and advanced sensors connected with Internet producing essential data on real-time. it helps in achieving three key outcomes of any efficient health care services- improved access to care, increased care quality, reduced care costs.

Smart Grid: Much like the Internet of Things as a whole, a smart grid is about balance. It is about efficiency.

It is about dynamically adjusting and re-adjusting to optimally deliver energy at the lowest cost and highest quality possible. A smart grid has the net effect of offering consumers the ability to participate in the solution

Green Devices

Environmentally friendly components contributing to a low-carbon society through size and weight reductions and original materials.

Here are some examples of Green Devices that give special consideration to the environment, such as saving energy and resources, and which were developed under the "Green Device Guidelines."

Super Green Device Examples

Super Green Devices are defined as devices that offer exceptional environmental performance.

These devices must be either the industry's No.1 devices, or the industry's first devices in at least one item of the External Environmental Claim Standards. In addition, they must satisfy at least 95% of the 21 assessment items (including 10 compulsory items) listed in Sharp's Environmental Performance Criteria. Some examples of Super Green Devices are shown here.

5-M pixel CCD Camera Module LZ0P3770.

5-Mpixel camera module for mobile devices. Incorporates functions comparable to digital cameras, such as an optical 3x zoom and auto-focus. Moreover uses approximately 20% less power than previous models (1 megapixel conversion) and is compact enough (22.0 x 11.3 x 21.7 mm) to be easily incorporated into mobile devices.

DBS Front end Unit

At just 30.6 x 25.0 x 11.7 mm in size, this compact DBS front-end unit offers a low power consumption of 0.38 mW.

IR Detecting Unit for Remote Control

An electromagnetic shield is applied to the unit by adding a secondary molding using conductive resin. This new structure, made possible by double resin molding technology, eliminates the need for metal shielding used in conventional products. Power consumption has been reduced by 60% and the unit is 33% lighter than previous models.

CCD Camera Module Power-supply IC

This power supply IC integrates the switching power supply and charge-pump power supply into one chip. A high-efficiency design and a reduction of peripheral components enable a significantly smaller mounting area. Five Green Devices Every Home Should Have.

You don't have to install huge solar panels on your home or drive a hybrid vehicle to go green. Sometimes, the best steps to take are the littlest ones. With that in mind, here are five green devices that should be in every home, all of which are easy to find, and none of which will break the bank.

1. **A battery charger.** Too many of our devices and tools still need batteries, which end up in landfills and leach toxic chemicals into the ground. Stop the cycle by making the switch to rechargeable batteries. They'll save you money, reduce your waste, and make a huge difference.

2. **A solar-powered calculator**. Why use batteries at all if you can choose a device that doesn't need them?.

3. **A solar-powered charger.** For the devices that don't take your conventional AA batteries, try picking up a solar charger. You can use the device to charge your cell phone, MP3 player and other portable devices. Several are on the market, with prices starting under $100.

4. **Solar-powered LED flashlights**. These great new devices use both solar cells for power and low-energy LED bulbs to provide bright light. I picked one up at my local drug store last week for 15 bucks, and it's the best buy I've made in a long time.

5. **An ENERGY STAR**-qualified fill-in-the-blank. Just about any appliance you want to buy comes in two flavors: an energy hog, and an energy-efficient version. Look for that ENERGY STAR logo whenever you buy an appliance or piece of electronics. You'll use at least 25% less energy, which will save you plenty of green.

6.4.   E-Waste

"E-waste" is a popular, informal name for electronic products nearing the end of their "useful life."E-wastes are considered dangerous, as certain components of some electronic products contain materials that are hazardous, depending on their condition and density. The hazardous content of these materials pose a threat to human health and environment. Discarded computers, televisions, VCRs, stereos, copiers, fax machines, electric lamps, cell phones, audio equipment and batteries if improperly disposed can leach lead and other substances into soil and groundwater. Many of these products can be reused, refurbished, or recycled in an environmentally sound manner so that they are less harmful to the ecosystem.

Industrial revolution followed by the advances in information technology during the last century has radically changed people's lifestyle. Although this development has helped the human race, mismanagement has led to new problems of contamination and pollution.

The technical prowess acquired during the last century has posed a new challenge in the management of wastes. For example, personal computers (PCs) contain certain components, which are highly toxic, such as chlorinated and brominated substances, toxic gases, toxic metals, biologically active materials, acids, plastics and plastic additives. The hazardous content of these materials pose an environmental and health threat. Thus proper management is necessary while disposing or recycling e wastes.

These days computer has become most common and widely used gadget in all kinds of activities ranging from schools, residences, offices to manufacturing industries. E-toxic components in computers could be summarized as circuit boards containing heavy metals like lead & cadmium; batteries containing cadmium; cathode ray tubes with lead oxide & barium; brominates flame retardants used on printed circuit boards, cables and plastic casing; poly vinyl chloride (PVC) coated copper cables and plastic computer casings that release highly toxic dioxins & furans when burnt to recover valuable metals; mercury switches; mercury in flat screens; poly chlorinated biphenyl's (PCB's) present in older capacitors; transformers; etc. Basel Action Network (BAN) estimates that the 500 million computers in the world contain 2.87 billion kgs of plastics, 716.7 million kgs of lead and 286,700 kgs of mercury. The average 14-inch monitor uses a tube that contains an estimated 2.5 to 4 kgs of lead. The lead can seep into the ground water from landfills thereby contaminating it. If the tube is crushed and burned, it emits toxic fumes into the air.

Effects on Environment and Human Health

Disposal of e-wastes is a particular problem faced in many regions across the globe. Computer wastes that are landfilled produces contaminated leachates which eventually pollute the groundwater. Acids and sludge obtained from melting computer chips, if disposed on the ground causes acidification of soil. For example, Guiyu, Hong Kong a thriving area of illegal e-waste recycling is facing acute water shortages due to the contamination of water resources.

This is due to disposal of recycling wastes such as acids, sludges etc. in rivers. Now water is being transported from faraway towns to cater to the demands of the population. Incineration of e-wastes can emit toxic fumes and gases, thereby polluting the surrounding air. Improperly monitored landfills can cause environmental hazards. Mercury will leach when certain electronic devices, such as circuit breakers are destroyed. The same is true for polychlorinated biphenyls (PCBs) from condensers.

When brominated flame retardant plastic or cadmium containing plastics are landfilled, both polybrominated dlphenyl ethers (PBDE) and cadmium may leach into the soil and groundwater. It has been found that significant amounts of lead ion are dissolved from broken lead containing glass, such as the cone glass of cathode ray tubes, gets mixed with acid waters and are a common occurrence in landfills.

Not only does the leaching of mercury poses specific problems, the vaporization of metallic mercury and dimethylene mercury, both part of Waste Electrical and Electronic Equipment (WEEE) is also of concern.

In addition, uncontrolled fires may arise at landfills and this could be a frequent occurrence in many countries.

When exposed to fire, metals and other chemical substances, such as the extremely toxic dioxins and furans (TCDD tetrachloro dibenzo-dioxin, PCDDs-polychlorinated dibenzodioxins. PBDDs-polybrominated dibenzo-dioxin and PCDFspoly chlorinated dibenzo furans) from halogenated flame retardant products and PCB containing condensers can be emitted. The most dangerous form of burning e-waste is the open-air burning of plastics in order to recover copper and other metals.

The toxic fall-out from open air burning affects both the local environment and broader global air currents, depositing highly toxic by products in many places throughout the world.

Table I summarizes the health effects of certain constituents in e-wastes. If these electronic items are discarded with other household garbage, the toxics pose a threat to both health and vital components of the ecosystem.

In view of the ill-effects of hazardous wastes to both environment and health, several countries exhorted the need for a global agreement to address the problems and challenges posed by hazardous waste.

Also, in the late 1980s, a tightening of environmental regulations in industrialized countries led to a dramatic rise in the cost of hazardous waste disposal. Searching for cheaper ways to get rid of the wastes, "toxic traders" began shipping hazardous waste to developing countries.

International outrage following these irresponsible activities led to the drafting and adoption of strategic plans and regulations at the Basel Convention. The Convention secretariat, in Geneva, Switzerland, facilitates and implementation of the Convention and related agreements. It also provides assistance and guidelines on legal and technical issues, gathers statistical data, and conducts training on the proper management of hazardous waste.

Basel Convention

The fundamental aims of the Basel Convention are the control and reduction of transboundary movements of hazardous and other wastes including the prevention and minimization of their generation, the environmentally sound management of such wastes and the active promotion of the transfer and use of technologies.

A Draft Strategic Plan has been proposed for the implementation of the Basel Convention. The Draft Strategic Plan takes into account existing regional plans, programmes or strategies, the decisions of the Conference of the Parties and its subsidiary bodies, ongoing project activities and process of international environmental governance and sustainable development.

The Draft requires action at all levels of society: training, information, communication, methodological tools, capacity building with financial support, transfer of know-how, knowledge and sound, proven cleaner technologies and processes to assist in the concrete implementation of the Basel Declaration. It also calls for the effective involvement and coordination by all concerned stakeholders as essential for achieving the aims of the Basel Declaration within the approach of common but differentiated responsibility.

Table 1: Effects of E-Waste Constituent on Health

Source of e-wastes	Constituent	Health effects
Solder in printed circuit boards, glass panels and gaskets in computer monitors	Lead (PB)	• Damage to central and peripheral nervous systems, blood systems and kidney damage. • Affects brain development of children.
Chip resistors and semiconductors	Cadmium (CD)	• Toxic irreversible effects on human health. • Accumulates in kidney and liver. • Causes neural damage. • Teratogenic.
Relays and switches, printed circuit boards	Mercury (Hg)	• Chronic damage to the brain. • Respiratory and skin disorders due to bioaccumulation in fishes.
Corrosion protection of untreated and galvanized steel plates, decorator or hardner for steel housings	Hexavalent chromium (Cr) VI	• Asthmatic bronchitis. • DNA damage.
Cabling and computer housing	Plastics including PVC	Burning produces dioxin. It causes • Reproductive and developmental problems; • Immune system damage; • Interfere with regulatory hormones
Plastic housing of electronic equipments and circuit boards.	Brominated flame retardants (BFR)	• Disrupts endocrine system functions
Front panel of CRTs	Barium (Ba)	Short term exposure causes: • Muscle weakness; • Damage to heart, liver and spleen.
Motherboard	Beryllium (Be)	• Carcinogenic (lung cancer) • Inhalation of fumes and dust. Causes chronic beryllium disease or beryllicosis. • Skin diseases such as warts.

Management of E-Wastes

It is estimated that 75% of electronic items are stored due to uncertainty of how to manage it. These electronic junks lie unattended in houses, offices, warehouses etc. and normally mixed with household wastes, which are finally disposed off at landfills. This necessitates implementable management measures. In industries management of e-waste should begin at the point of generation. This can be done by waste minimization techniques and by sustainable product design. Waste minimization in industries involves adopting:

- Inventory management.
- Production-process modification.
- Volume reduction.
- Recovery and reuse.

Inventory management

Proper control over the materials used in the manufacturing process is an important way to reduce waste generation (Freeman, 1989). By reducing both the quantity of hazardous materials used in the process and the amount of excess raw materials in stock, the quantity of waste generated can be reduced. This can be done in two ways i.e. establishing material-purchase review and control procedures and inventory tracking system.

Developing review procedures for all material purchased is the first step in establishing an inventory management program. Procedures should require that all materials be approved prior to purchase. In the approval process all production materials are evaluated to examine if they contain hazardous constituents and whether alternative non-hazardous materials are available.

Another inventory management procedure for waste reduction is to ensure that only the needed quantity of a material is ordered. This will require the establishment of a strict inventory tracking system. Purchase procedures must be implemented which ensure that materials are ordered only on an as-needed basis and that only the amount needed for a specific period of time is ordered.

Production-Process Modification

Changes can be made in the production process, which will reduce waste generation. This reduction can be accomplished by changing the materials used to make the product or by the more efficient use of input materials in the production process or both.

Potential waste minimization techniques can be broken down into three categories:

1. Improved operating and maintenance procedures.
2. Material change.
3. Process-equipment modification.

Improvements in the operation and maintenance of process equipment can result in significant waste reduction. This can be accomplished by reviewing current operational procedures or lack of procedures and examination of the production process for ways to improve its efficiency. Instituting standard operation procedures can optimise the use of raw materials in the production process and reduce the potential for materials to be lost through leaks and spills. A strict maintenance program, which stresses corrective maintenance, can reduce waste generation caused by equipment failure. An employee-training program is a key element of any waste reduction program. Training should include correct operating and handling procedures, proper equipment use, recommended maintenance and inspection schedules, correct process control specifications and proper management of waste materials.

Hazardous materials used in either a product formulation or a production process may be replaced with a less hazardous or non-hazardous material. This is a very widely used technique and is applicable to most manufacturing processes. Implementation of this waste - reduction technique may require only some minor process adjustments or it may require extensive new process equipment. For example, a circuit board manufacturer can replace solvent-based product with water-based flux and simultaneously replace solventvapor degreaser with detergent parts washer.

Installing more efficient process equipment or modifying existing equipment to take advantage of better production techniques can significantly reduce waste generation. New or updated equipment can use process materials more efficiently producing less waste. Additionally such efficiency reduces the number of rejected or off-specification products, thereby reducing the amount of material which has to be reworked or disposed of. Modifying existing process equipment can be a very cost-effective method of reducing waste generation. In many cases the modification can just be relatively simple changes in the way the materials are handled within the process to ensure that they are not wasted. For example, in many electronic manufacturing operations, which involve coating a product, such as electroplating or painting, chemicals are used to strip off coating from rejected products so that they can be recoated. These chemicals, which can include acids, caustics, cyanides etc are often a hazardous waste and must be properly managed. By reducing the number of parts that have to be reworked, the quantity of waste can be significantly reduced.

Volume Reduction

Volume reduction includes those techniques that remove the hazardous portion of a waste from a non-hazardous portion. These techniques are usually to reduce the volume, and thus the cost of disposing of a waste material. The techniques that can be used to reduce waste-stream volume can be divided into 2 general categories: source segregation and waste concentration. Segregation of wastes is in many cases a simple and economical technique for waste reduction. Wastes containing different types of metals can be treated separately so that the metal value in the sludge can be recovered. Concentration of a waste stream may increase the likelihood that the material can be recycled or reused. Methods include gravity and vacuum filtration, ultra filtration, reverse osmosis, freeze vaporization etc.

For example, an electronic component manufacturer can use compaction equipments to reduce volume of waste cathode ray-tube.

Recovery and Reuse

This technique could eliminate waste disposal costs, reduce raw material costs and provide income from a salable waste. Waste can be recovered on-site, or at an off-site recovery facility, or through inter industry exchange. A number of physical and chemical techniques are available to reclaim a waste material such as reverse osmosis, electrolysis, condensation, electrolytic recovery, filtration, centrifugation etc. For example, a printed-circuit board manufacturer can use electrolytic recovery to reclaim metals from copper and tin-lead plating bath. However recycling of hazardous products has little environmental benefit if it simply moves the hazards into secondary products that eventually have to be disposed of. Unless the goal is to redesign the product to use nonhazardous materials, such recycling is a false solution.

Sustainable Product Design

Minimization of hazardous wastes should be at product design stage itself keeping in mind the following factors*

- *Rethink the product design:* Efforts should be made to design a product with fewer amounts of hazardous materials. For example, the efforts to reduce material use are reflected in some new computer designs that are flatter, lighter and more integrated. Other companies propose centralized networks similar to the telephone system.

- *Use of renewable materials and energy:* Bio-based plastics are plastics made with plant-based chemicals or plant-produced polymers rather than from petrochemicals.

Bio-based toners, glues and inks are used more frequently. Solar computers also exist but they are currently very expensive.

- ***Use of non-renewable materials that are safer:*** Because many of the materials used are non-renewable, designers could ensure the product is built for re-use, repair and/or upgradeability. Some computer manufacturers such as Dell and Gateway lease out their products thereby ensuring they get them back to further upgrade and lease out again.

Management Options

Considering the severity of the problem, it is imperative that certain management options be adopted to handle the bulk e-wastes. Following are some of the management options suggested for the government, industries and the public.

Responsibilities of the Government

1. Governments should set up regulatory agencies in each district, which are vested with the responsibility of co-ordinating and consolidating the regulatory functions of the various government authorities regarding hazardous substances.

2. Governments should be responsible for providing an adequate system of laws, controls and administrative procedures for hazardous waste management (Third World Network. 1991). Existing laws concerning e-waste disposal be reviewed and revamped. A comprehensive law that provides e-waste regulation and management and proper disposal of hazardous wastes is required. Such a law should empower the agency to control, supervise and regulate the relevant activities of government departments. Under this law, the agency concerned should.

    - Collect basic information on the materials from manufacturers, processors and importers and to maintain an inventory of these materials. The information should include toxicity and potential harmful effects.

    - Identify potentially harmful substances and require the industry to test them for adverse health and environmental effects.

    - Control risks from manufacture, processing, distribution, use and disposal of electronic wastes.

    - Encourage beneficial reuse of "e-waste" and encouraging business activities that use waste". Set up programs so as to promote recycling among citizens and businesses.

    - Educate e-waste generators on reuse/recycling options.

3. Governments must encourage research into the development and standard of hazardous waste management, environmental monitoring and the regulation of hazardous waste-disposal.

4. Governments should enforce strict regulations against dumping e-waste in the country by outsiders. Where the laws are flouted, stringent penalties must be imposed. In particular, custodial sentences should be preferred to paltry fines, which these outsiders / foreign nationals can pay.

5. Governments should enforce strict regulations and heavy fines levied on industries, which do not practice waste prevention and recovery in the production facilities.

6. Polluter pays principle and extended producer responsibility should be adopted.

7. Governments should encourage and support NGOs and other organizations to involve actively in solving the nation's e-waste problems.

8. Uncontrolled dumping is an unsatisfactory method for disposal of hazardous waste and should be phased out.

9. Governments should explore opportunities to partner with manufacturers and retailers to provide recycling services.

Responsibility and Role of Industries

1. Generators of wastes should take responsibility to determine the output characteristics of wastes and if hazardous, should provide management options.

2. All personnel involved in handling e-waste in industries including those at the policy, management, control and operational levels, should be properly qualified and trained. Companies can adopt their own policies while handling e-wastes. Some are given below:

    - Use label materials to assist in recycling (particularly plastics).
    - Standardize components for easy disassembly.
    - Re-evaluate 'cheap products' use, make product cycle 'cheap' and so that it has no inherent value that would encourage a recycling infrastructure.
    - Create computer components and peripherals of biodegradable materials.
    - Utilize technology sharing particularly for manufacturing and de manufacturing.
    - Encourage/promote/require green procurement for corporate buyers.
    - Look at green packaging options.

3. Companies can and should adopt waste minimization techniques, which will make a significant reduction in the quantity of e-waste generated and thereby lessening the impact on the environment. It is a "reverse production" system that designs

infrastructure to recover and reuse every material contained within e-wastes metals such as lead, copper, aluminum and gold, and various plastics, glass and wire. Such a "closed loop" manufacturing and recovery system offers a win-win situation for everyone, less of the Earth will be mined for raw materials, and groundwater will be protected, researchers explain.

4. Manufacturers, distributors, and retailers should undertake the responsibility of recycling/disposal of their own products.

5. Manufacturers of computer monitors, television sets and other electronic devices containing hazardous materials must be responsible for educating consumers and the general public regarding the potential threat to public health and the environment posed by their products. At minimum, all computer monitors, television sets and other electronic devices containing hazardous materials must be clearly labeled to identify environmental hazards and proper materials management.

Responsibilities of the Citizen

Waste prevention is perhaps more preferred to any other waste management option including recycling. Donating electronics for reuse extends the lives of valuable products and keeps them out of the waste management system for a longer time. But care should be taken while donating such items i.e. the items should be in working condition.

Reuse, in addition to being an environmentally preferable alternative, also benefits society. By donating used electronics, schools, non-profit organizations, and lower-income families can afford to use equipment that they otherwise could not afford. E-wastes should never be disposed with garbage and other household wastes. This should be segregated at the site and sold or donated to various organizations. While buying electronic products opt for those that:

- Are made with fewer toxic constituents.
- Use recycled content.
- Are energy efficient.
- Are designed for easy upgrading or disassembly.
- Utilize minimal packaging.
- Offer leasing or take back options.
- Have been certified by regulatory authorities. Customers should opt for upgrading their computers or other electronic items to the latest versions rather than buying new equipments.

NGOs should adopt a participatory approach in management of e-wastes.

6.5. E-waste Recycling

E-waste recycling is the reuse and reprocessing of electrical and electronic equipment of any type that has been discarded or regarded as obsolete. Some of the common E-wastes include: home appliances such as televisions, air conditioners, electric cookers and heaters, air condoners, fans, DVDs, Radios and microwaves among others; information technology equipments such as computers, mobile phones, laptops, batteries, circuit boards, hard disks, and monitors among others; and other electronic utilities such as leisure, lighting, and sporting equipments. Recycling of e-waste is a growing trend and was initiated to protect human and environmental health mainly due to the widespread environmental pollution impacts of e-waste.

Why Is Electronics Recycling Important?

- **Rich Source of Raw Materials** Internationally, only 10-15 percent of the gold in e-waste is successfully recovered while the rest is lost. Ironically, electronic waste contains deposits of precious metal estimated to be between 40 and 50 times richer than ores mined from the earth, according to the United Nations.
- **Solid Waste Management** Because the explosion of growth in the electronics industry, combined with short product life cycle has led to a rapid escalation in the generation of solid waste.
- **Toxic Materials** Because old electronic devices contain toxic substances such as lead, mercury, cadmium and chromium, proper processing is essential to ensure that these materials are not released into the environment. They may also contain other heavy metals and potentially toxic chemical flame retardants.
- **International Movement of Hazardous Waste** The uncontrolled movement of e-waste to countries where cheap labor and primitive approaches to recycling have resulted in health risks to local residents exposed to the release of toxins continues to an issue of concern.

Step-by Step Process of E-waste Recycling

The e-waste recycling process is highly labor intensive and goes through several steps. Below is the step-by-step process of how e-waste is recycled,

1. Picking Shed

When the e-waste items arrive at the recycling plants, the first step involves sorting all the items manually. Batteries are removed for quality check.

2. Disassembly

After sorting by hand, the second step involves a serious labor intensive process of manual dismantling. The e-waste items are taken apart to retrieve all the parts and then categorized into core materials and components. The dismantled items are then separated into various categories into parts that can be re-used or still continue the recycling processes.

3. First Size Reduction Process

Here, items that cannot be dismantled efficiently are shredded together with the other dismantled parts to pieces less than 2 inches in diameter. It is done in preparation for further categorization of the finer e-waste pieces.

4. Second Size Reduction Process

The finer e-waste particles are then evenly spread out through an automated shaking process on a conveyor belt. The well spread out e-waste pieces are then broken down further. At this stage, any dust is extracted and discarded in a way that does not degrade the environmentally.

5. Over-Band Magnet

At this step, over-band magnet is used to remove all the magnetic materials including steel and iron from the e-waste debris.

6. Non-Metallic and Metallic Components Separation

The sixth step is the separation of metals and non-metallic components. Copper, aluminum, and brass are separated from the debris to only leave behind non-metallic materials. The metals are either sold as raw materials or re-used for fresh manufacture.

7. Water Separation

As the last step, plastic content is separated from glass by use of water. One separated, all the materials retrieved can then be resold as raw materials for re-use. The products sold include plastic, glass, copper, iron, steel, shredded circuit boards, and valuable metal mix.

E-cycle Components Re-Use

1. **Plastic.** All the plastic materials retrieved are sent to recyclers who use them to manufacture items such as fence posts, plastic sleepers, plastic trays, vineyard stakes, and equipment holders or insulators among other plastic products.

2. **Metal**. Scrap metals materials retrieved are sent to recyclers to manufacture new steel and other metallic materials.

3. **Glass**. Glass is retrieved from the Cathode Ray Tubes (CRTs) mostly found in televisions and computer monitors. Extracting glass for recycling from CRTs is a more complicated task since CRTs are composed of several hazardous materials. Lead is the most dangerous and can adversely harm human health and the environment. Tubes in big CRT monitors can contain high levels of lead of up to 4 kilograms. Other toxic metals such as barium and phosphor are also contained in CRT tubes. To achieve the best environmentally friendly glass extraction, the following steps ensure a specialized CRT recycling:

   - Manual separation of the CRT from the television or monitor body
   - Size reduction process where the CRT is shredded into smaller pieces. Dust is eliminated and disposed in an environmentally friendly way.
   - All metals are removal through over-band magnets, where ferrous and non-ferrous components are eliminated from the glass materials.
   - A washing line is then used to clear oxides and phosphors from the glass
   - Glass sorting is the final step whereby leaded glass is separated from non-leaded glass. The extracts can then be used for making new screens.

4. **Mercury**. Mercury containing devices are sent to mercury recycling facilities that uses a specialized technology for elimination for use in dental amalgams and metric instruments, and for fluorescent lighting. Other components such as glass and plastics are re-used for manufacture of their respective products.

5. **Printed Circuit Boards**. Circuit boards are sent to specialized and accredited companies where they are smelted to recover non-renewable resources such as silver, tin, gold, palladium, copper and other valuable metals.

6. **Hard Drives**. Hard drives are shredded in whole and processed into aluminum ingots for use in automotive industry.

7. **Ink and Toner Cartridges**. Ink and toner cartridges are taken back to respective manufacturing industries for recycling. They are remanufactured while those that can't are separated into metal and plastic for re-use as raw materials.

8. **Batteries**. Batteries are taken to specialized recyclers where they are hulled to take out plastic. The metals are smelted is specialized conditions to recover nickel, steel, cadmium and cobalt that are re-used for new battery production and fabrication of stainless steel.

Trends in Electronic Waste Recycling

In the 1990s some European countries banned the disposal of electronic waste in landfills. This created an e-waste processing industry in Europe. Early in 2003 the EU presented the WEEE and RoHS directives for implementation in 2005 and 2006.

Some states in the U.S. developed policies banning CRTs from landfills. Some e-waste processing is carried out within the U.S. The processing may be dismantling into metals, plastics and circuit boards or shredding of whole appliances. From 2004 the state of California introduced a Electronic Waste Recycling Fee on all new monitors and televisions sold to cover the cost of recycling. The amount of the fee depends on the size of the monitor. That amount was adjusted on July 1, 2005 in order to match the real cost of recycling.

A typical electronic waste recycling plant as found in some industrialized countries combines the best of dismantling for component recovery with increased capacity to process large amounts of electronic waste in a cost effective-manner. Material is fed into a hopper, which travels up a conveyor and is dropped into the mechanical separator, which is followed by a number of screening and granulating machines. The entire recycling machinery is enclosed and employs a dust collection system. The European Union, South Korea, Japanand Taiwan have already demanded that sellers and manufacturers of electronics be responsible for recycling 75 percent of them.

Many Asian countries have legislated, or will do so, for electronic waste recycling.

The United States Congress is considering a number of electronic waste bills including the National Computer Recycling Act introduced by Congressman Mike Thompson (D-CA). This bill has continually stalled, however.

In the meantime, several states have passed their own laws regarding electronic waste management. California was the first state to enact such legislation, followed by Maryland, Maine, and Washington.

Green Engineering

Overview

Over the past year, the mainstream media has dramatically increased its emphasis on all things "green." Concerns about global climate change, soaring energy prices, and increased government legislation are driving new priorities and expectations – from consumer products to corporate responsibility and sustainability plans. To meet these new demands, companies, big and small, around the world are scrambling to not only create products and technologies that address these concerns but also change the ways and processes by which they are developed. Engineers and scientists worldwide are leading the charge to address one of the largest challenges society faces, and they have the unique opportunity to make a bigger impact on the environment than any government policy. Green engineering provides the tools, techniques, and technologies to foster this innovation.

1. What is Green Engineering?

Engineers who want to lower the emissions of their products, develop devices that consume less energy, create viable renewable energy technologies, or better understand the global ecosystem need green engineering.

Green engineering is the use of measurement and control techniques to design, develop, and improve products, technologies, and processes that result in environmental and economic benefits. While green may be the focus today, performing green engineering is fundamentally no different than any other type of engineering innovation.

First, you need to measure the variables with which you are concerned, and then you can begin the process of designing or "fixing" products and processes that achieve your desired goal.

Green engineering encompasses common measurements such as power quality and consumption; emissions from vehicles and factories, such as mercury and nitrogen oxides; and environmental data, including carbon, temperature, and water quality.

National Instruments enables green engineering by providing measurement, automation, and design tools that empower engineers and scientists to first quantify and understand real-world data and, second, correct problems by designing and developing the next generation of products and technologies with improved efficiency and reduced environmental impact.

2. Green Engineering Technology and Business Opportunity

Among the top concerns of green engineering are two questions: Is today's technology viable enough to make an impact, and is there a real business opportunity? Brief answers to these questions follow with case studies further illustrating green engineering.

Green engineering applications span almost every industry and range from monitoring the health of forests, so ecologists can better understand the effects of global warming, to retrofitting aging production facilities and machines with new control systems to make them more efficient.

Technology

The technology components required for green engineering are not only accessible but also easier to use and available at a lower price than ever before.

Some of the key technologies that enable green engineering include the following:

- Graphical software to measure and fix.
- High-speed and high-resolution measurements.
- Domain-specific analysis libraries.
- FPGAs for advanced control.

Some of these new technologies have resulted from growth in the semiconductor industry. This growth has created major advancements in the capabilities of analog-to-digital converters (ADCs) while mass adoption of consumer electronics has decreased cost. Other technologies have been around for some time, but new improvements to design and engineering tools have made them more usable by domain experts rather than solely technology experts. This shift puts the necessary technology directly into the hands of those who are closest to the problems, so they can develop solutions much more successfully than in the past.

Business Opportunity

There is a huge opportunity for profit and savings in green products. With oil prices surging to all-time highs, demand continues to be strong for technologies and products that help companies use less oil in their machines and processes while achieving the same output. Other companies, looking to avoid steep fines for nonadherence to environmental regulations, are buying monitoring and reporting tools. Furthering this trend, a recent study by PricewaterhouseCoopers found that venture capital investment in clean technology applications, such as energy conservation, recycling, water purification, emissions control, and renewable energy, tripled in 2006 to more than $1.4 billion USD and grew again in 2007 at nearly 50 percent to more than $2 billion USD.

3. Green Engineering Applied

Green engineering applications span almost every industry and range from monitoring the health of forests, so ecologists can better understand the effects of global warming, to retrofitting aging production facilities and machines with new control systems to make them more efficient.

While there are many ways to group these green applications, most fall into the following five categories:

1. Renewable power generation.

2. Power quality.

3. Environmental monitoring.

4. Machine and process optimization.

5. Development and test of green products and technologies.

The following examples demonstrate green engineering in renewable power generation and machine and process optimization.

Renewable Power Generation

Renewable power generation covers a wide range of technologies including wind, solar (photovoltaic and thermal), bio fuels, hydro, wave harvesting, geothermal, and even high-energy physics. Research and development in these areas are exploding around the world, driven by environmental suitability goals and ever-increasing government legislation. Today more than 50 countries, with a variety of political, geographical, and economic conditions, have set aggressive targets for the amount of energy generated from renewable sources.

State/Country	Renewable Energy Goal (As a Percent of Total Energy)	Target Year
Texas	10%	2025
California	33%	2020
United Kingdom	10%	2010
France	20%	2020
Sweden	60%	2020
China (Phase II)	15%	2020
European Union	20%	2020

This table lists a few examples of the targets various governments have set for renewable energy goals. With mandates of up to 60 percent and deadlines as close as 2010, the innovation and resources dedicated to reaching these goals are significant engineering challenges. To put this into perspective, only 3 percent of the energy consumed worldwide in 2007 was from renewable sources. While this may seem daunting with much work to be done, even the last two years have shown significant progress toward these goals. Engineers and scientists have historically risen to meet seemingly far-fetched challenges, such as putting a man on the moon, and what makes today's situation even more hopeful is the global scope. To meet these goals, engineers are scrambling to design new technologies, and many are using NI tools to meet tight deadlines and complex specifications.

Wind power technologies create an incredible variety of challenges for engineers developing and validating new designs. One of the biggest challenges is developing accurate control systems to reduce damage on turbine components caused by high winds. These engineers must use complex algorithms to adjust the pitch of the blades to maintain a constant rotating speed in variable wind conditions. NI LabVIEW Real-Time software and PXI hardware are key components in prototyping these algorithms, testing their reliability, and validating their performance. Additionally, wind turbine engineers also need to design for increasingly sophisticated structural dynamics as bigger blades, some up to 350 ft, are installed to generate larger amounts of electricity.

Solar power manufacturers also face significant engineering obstacles to lower the material costs of solar cells and increase their production efficiency. They need simpler, faster ways to perform photovoltaic device output performance tests, such as current-voltage (I-V) characterization; detailed, precise control for the semiconductor fabrication process; and accurate power quality measurements of the inverters that connect solar arrays to the grid.

Although by no means exhaustive, the requirements of wind and solar power applications echo the needs of all engineers developing renewable energy applications for better ways to measure and fix their next-generation technologies. With LabVIEW providing a common platform for instrumentation, control systems, prototyping, and validation, engineers can more rapidly iterate on their designs, getting new technologies to market quickly and economically.

Machine and Process Optimization

When Nucor Steel, one of the largest steel companies in the world and America's largest recycler, acquired the Marion Steel Company in 2005, one of its first actions was to add automation systems throughout the newly acquired Marion, Ohio, minimill plant to increase efficiency and safety. Last year alone, Nucor recycled more than 22 million tons of steel, including 9 million cars. The process of melting and recasting steel requires a large amount of electricity, and even small increases in efficiency throughout this process result in huge energy and economic savings.

Dave Brandt, an electrical engineer at Nucor Steel Marion Inc., was charged with implementing the automation systems. Brandt used NI tools, including programmable automation controllers (PACs) and LabVIEW, to develop a variety of automation systems such as a scale and weighing system, an online reactor in series with the furnace, and a remote switching station, which have greatly reduced electricity usage, eliminated potential safety issues, and contributed to Nucor's pioneering commitment to environmental stewardship.

Brandt used LabVIEW and NI Compact Field Point hardware to create a scale and weighing system to know the exact amount of steel and, therefore, the exact amount of energy needed to heat its electricity-powered furnace. Before Nucor implemented this system, the company estimated the amount of steel in each burn, which resulted in "hit or miss" results and oftentimes overheated the steel, wasting electricity in the process and producing unacceptable-quality newly cast steel. As a result, the steel would have to be reheated, which used a significant amount of energy and cost Nucor a lot of money. Since implementing this weighing system, Nucor has drastically decreased the amount of reheats it performs, reducing the 2007 total number to 10 out of more than 6,000 batches.

QUESTION BANK

Part A (2 Marks)

Unit I: Natural Resources

1. Define Environmental science.

   Environmental science is the study of physical, chemical, and also biological conditions, which surround the 'living beings', who influence them directly or indirectly.

2. What are the important components of environment?

   Abiotic or non-living components Biotic or living component Energy component

3. What are the processes involved in hydrological cycle?

   Continuous evaporation, transpiration, precipitation of surface run off and ground water.

4. Define biogeochemical cycle. Give example.

   The continuous circulation of all the essential elements and compounds required for life, from the environment to the organism and back to the environment. e.g., carbon cycle.

5. What are the functions of lithosphere?

   It is a home for human beings and wild lives. It is a store house minerals and organic matters.

6. Explain biosphere?

   The part of lithosphere, hydrosphere and atmosphere in which living organisms live and interact with one another is called biosphere.

7. Write any two adverse effects caused by overgrazing.

   - Land degradation.
   - Loss of useful species.

8. What is desertification? Give any two reasons for it.

   Desertification means degradation of one fertile land to desert like land. Reason deforestation, overgrazing, mining, overgrazing.

9. Define Land Degradation.

   Land degradation means process of deterioration of soil or loss of fertility of soil.

10. Distinguish renewable and non-renewable sources of energy.

Energy Sources	Advantage	Disadvantage
Renewable	Wide availability	Unreliable supply
	Low cost	Produced in small quantity
	Decentralized power production	Difficult to store
		Cost more
	Low pollution	
	Available for the future	
Non Renewable	Available in high concentrated form	highly pollution
		Available only in few places
	Easy to store	High running cost
	Reliable supply Lower cost	Limited supply and will one day get exhausted

11. What are the reasons for deforestation?

Deforestation means increasing agricultural production, increasing industrial activity, increasing demands for wood resources.

12. What is an aquifer?

A highly permeable layer of sediment or rock containing water.

13. What does strategic metals and minerals mean?

These are the metals and minerals that a country uses but cannot produce itself. Essential for defence. e.g., cobalt, iron, manganese.

14. What is water logging and how it is prevented?

Saturation of soil with irrigation water or excessive precipitation so that the water table rises close to surface.

- Prevent excessive irrigation
- Subsurface drainage and bio drainage by trees like Eucalyptus trees are adopted.
- Leakage from water pump are detected.

15. Write any two effects of ground water depletion.

- Lower the surface water level.
- Land subsidence.
- Salt water intrusion.
- Climate change.

16. List any four adverse affects of mining.

- Scarring and disruption of land surface.
- Land subsidence.
- Smelting causes air pollution.
- Acid mine drainage contaminates ground water.

17. What is artesian well?

A well or hole in aquifer flows freely at the surface.

18. What is wetland? Give examples and use.

Wetlands are the natural water storage bodies on ground surface Eg. Swamps, Meadows, Marshes.

19. State the major process which have major environmental impact while processing of minerals.

Smelting, chemical extractions.

20. What is water salinity?

During over irrigation, all the water is not absorbed in the soil. such water evaporates leaving behind a thin crust of dissolved salts in the top soil.

21. Give two examples of primary and secondary sources of energy?

Primary source-fossil fuels, hydro energy Secondary source-petrol and electrical energy.

22. What is ocean thermal energy conversation?

In oceans, a thermal gradient (i.e., the temperature difference) of about 200C exists between surface water heated by sun and colder deep water. This difference can be harnessed to produce power. This concept is OTEC.

23. Give any four environmental benefits of dam.

Source of cleaner and safer power for irrigation of agricultural lands Helps in recharging of ground water Habitat for many fishes and wildlife

24. What are the sources of water?

Ground water, surface water, sea water, rain water.

25. What is sardar sarovar Narmada project?

It is a multipurpose project on river Narmada, bringing the benefits of irrigation, power and drinking water to Gujarat, Mathya Pradesh and Rajasthan.

26. What is integrated pest management?

The process of controlling crop pests using ecological system.

Unit II: Ecosystem and Biodiversity

1. Differentiate between a biome and an ecosystem.

On earth there are many sets of ecosystems which are exposed to same climatic conditions and having dominant species with similar lifecycle, Climatic adaptations and physical structure. This set of Ecosystem is called Biome (Small ecosystem) A group of Organisms Interacting among themselves and with environment is known as ecosystem. is the basic functional unit of Ecology.

2. Define: Food chain and Food Web.

In linear food chains, if one species gets affected or becomes extinct, then the species in the subsequent tropic levels are also affected. Network of food chain–Food web–if one species gets affected, it does not affect other tropic levels so seriously.

3. Differentiate between genetic diversity and species diversity.

Genetic diversity–diversity within species ie., variations of genes within the species. Species diversity-diversity between different species. The sum of varieties of all the living organisms at the species level is known as species diversity.

4. Define the terms producers and consumers

Producers – depend on their food themselves through Photosynthesis. Eg. All green plants , trees. Consumers. - depend directly or indirectly on the producers eg. Plant eating species, animals eating species.

5. What are ecological pyramids?

Graphical representation of structural and function of tropic levels of an ecosystem is called ecological pyramids.

6. Define "Hot spots of biodiversity".

The hot spots are the geographic areas which possess high endemic species.

7. Define biodiversity.

Biodiversity – the variety and variability among all groups of living organisms and the ecosystem in which they occur.

8. Define ecology.

Ecology – study of interactions among organisms or group of organisms with their environment (Biotic and Abiotic Organisms).

9. What is ecological succession?

The progressive replacement of one community by another till the development of stable community in a particular area is called ecological succession.

10. What are estuaries? Give its importance

Estuaries are bays or semi-enclosed bodies of brackish water that form where river enter the ocean. They are nursuries for economically important fish, crabs, shrimp, and oysters, etc.

11. Give any four biogeographically regions in India.

   1. Himalayan zone,
   2. Western Ghats,
   3. Gangetic plain and
   4. Andaman and Nicobar islands.

12. Name any six biosphere in India.

   1. Nilgiri
   2. Sundarbans
   3. Great Nicobar
   4. Gulf of mannar
   5. Kanchenjunga
   6. Agathyamalai.

13. What is endangered species? Give an example in India.

The species which are in the danger of extinction is called endangered species. Examples: Lion, Nigiri Tahr.

14. What are the threats to biodiversity?

   1. Habitat destruction
   2. Fragmentation
   3. Hunting and fishing

15. What do you mean by 'In-situ' conservation?

Conserving wildlife by allocating large portion of earth' surface.

Unit III: Environmental Pollution

1. Give any three methods of air pollution control equipment.

   Catalytic converters Cyclone collectors Electrostatic precipitators

2. What is smog?

   It is a mixture of smoke and fog which forms droplets that remain suspended in the air.

3. What is point source of water pollution?

   Point source is discharge pollutants at specific locations through pipes, ditches into bodies of surface waters.

4. When a sound causes noise pollution?

   Noise beyond 120 dB.

5. What are the types of solid waste?

   Municipal waste, Industrial waste, Hazardous waste.

6. What is waste minimization?

   Industrial manufacturing system the primary concern should be reducing the quantities of waste materials produced. This avoids the necessity to threat and disposal off such materials.

7. Name the chemical constituent of gas that caused death in Bhopal gas tragedy in India.

   Methyl iso cyanide.

8. Define the term Tsunami.

   It is a large wave that are generated in a water body where the sea floor deformed by seismic activity. This activity displaces the overlying water in the ocean.

9. What is marine pollution?

   Marine pollution is defined as the discharge of waste substances in to the resulting in harm to living resource, hazards to human health, hindrance to fishery and impairment of quality for use of sea water.

10. Define a) Decibel b) COD

    Decibel: Decibel (dB) is defined as the one tenth of the longest unit Bel.

    COD: COD (Chemical Oxygen Demand) is the amount oxygen required for chemical oxidation of organic matter using some oxidizing agent like K2Cr2O7 and KMnO4.

11. What are the effects of noise pollution?

Interferes with man's communication Hearing damage.

Physiological and psychological changes.

12. How solid wastes are disposed ultimately?

Landfill, incineration and composting.

13. What are the causes of water pollution?

Domestic sewage, Industrial effluents, Synthetic detergents, Agro chemicals, Oil, Thermal pollutants, Run off from land fills

14. What is disaster? Give few examples

Disaster is a geological processes and is defined as the sudden calamity which brings misfortune and miseries to human community e.g., flood, cyclone, landslide, earthquake and Tsunami.

15. Differentiate between primary and secondary air pollutants with examples:

Primary pollutants	Secondary pollutants
These are emitted directly in the Atmosphere	These are pollutants in which some of in harmful form the primary air pollutants may react with one another to form new pollutants.
Examples:CO,NO,SO2	Examples: NO,NO2------ HNO3/NO3

16. Differentiate between pollution prevention and pollution control

Pollution prevention	**Pollution control**
It means using processes, practices, materials, products or energy that avoid or minimize the creation of pollutants and waste or environmental disturbances and reduse risk to human health	The proper control measures practiced to minimize the pollution level

17. Give any six air pollutants

Sulphur dioxide, Nitrogen oxides, Sulphur trioxide, Carbon dioxide, Hydrocarbons, Carbon monoxide

18. What is the objective of Green chemistry?

To promote research and development, and implementation of innovative chemical technologies that accomplishes pollution preventions in scientifically sound and cost-effective manner.

Unit IV: Social Issues and Green Approaches

1. What is sustainable development?

   Sustainable development is defined as meeting the needs of the present without compromising the ability of future generations to meet their own needs, or extending progress, without exhausting resources, beyond the foreseeable future.

2. Define watershed management.

   The management of rainfall and resultant run off is called watershed management.

3. What is green house effect?

   Green house effect may be defined as the progressive warming up of the earth's surface due to blanketing effect of manmade carbondioxide in the atmosphere.

4. What are the advantages of rainwater harvesting?

   Reduction in the use of current for pumping water.

   Mitigation by effects of droughts and achieving drought proofing.

   Increasing the availability of water from well, Rise in ground water level and minimizing the soil erosion and flood hazards.

5. What is acid rain?

   The presence of excessive acids in rain water is called as acid rain.

6. What are the effects of acid rain?

   Acid rain corrodes houses, monuments, statues, bridges and fences.

   Deteriorate the paint and stone.

7. What are the objectives of watershed management?

   To minimize the risk of foods, droughts and landslides.

   To develop rural areas in the region with clear plan for improving the economy.

   To generate huge employment opportunities.

   To protect the soil from erosion by runoff.

   To rise ground water level.

8. Define the term environmental ethics.

   Environmental ethics refers to the issues, principals and guidelines relating to human interactions with their environment.

9. What are the effects of global warming?

    Increase the sea level.

    Negative effect on crop production and forest growth Decrease the water resource.

    Increase the drought.

10. Explain the factors affecting watershed management.

    Overgrazing, deforestation, mining, construction activities degrades watershed.

11. What is meant by Environmental audit?

Environmental audits are indented to quantify environmental performance and environmental position.

12. Write the consequences of ozone layer depletion.

    Damage genetic materials in the skin cells which cause skin cancer. Affect the aquatic forms, Global warming, Degradation of paints, plastics.

13. What is waste land reclamation?

    The restoration of disturbed land to ecologically stable condition. To make the land more productive for agriculture.

14. What are the state enactments of environmental legislations of India?

    Smoke control, Landuse, Pest control, Water pollution.

Unit V: Human Population and Environment

1. Differentiate between HIV and AIDS.

    HIV-Human Innuno deficiency virus cause AIDS disease. Virus is passed through infected blood, semen. AIDS-Acquired Immuno Deficiency Syndrome Acquired means disease is not hereditary but develops after birth from contact with a disease causing agent. Immune deficiency means that the disease is characterized by a weakening of immune system.

2. What are the major precautions to avoid AIDS?

    Education, prevention of blood borne HIV transmission, primary health care, counseling services, drug treatment.

3. Define "Human rights".

    Human rights are the fundamental rights, which are possessed by all human being irrespective of their caste, nationality, sex and language.

4. What is population explosion?

    The enormous increase in population due to low death rate (mortalityt) and high birth rate (natality), is termed as population explosion.

5. Write any two applications of information technology in environment.

Land and water management, Information on type, density, biomass, forest fire, pest and disease.

6. What is meant by population doubling time? How it is calculated?

The number of years takes for doubling of population is called population doubling time.

7. What are the reasons responsible for population explosion?

Invention of modern medical facilities reduces the death rate and increases the birth rate.
Increase of life expectancy, Illiteracy.

8. Write the value of education to the society.

Improve the integral growth of human being.

Create attitude and improvement towards sustainable lifestyle.

To understand about natural environment.

9. What are the major objectives of family welfare programme in India?

Reduce infant mortality rate to below 30/100 infants.

Achieves 100% registration of birth, death and marriage.

Encourage late marriage and later child birth.

Constrain the spread of AIDS/HIV.

10. What do you mean by carrying capacity of population?

The maximum population size that an ecosystem can support under particular environmental condition is known as the carrying capacity of population.

11. What are the reasons for population explosion?

Improved sanitary conditions, Better health care, increase in productivity of agriculture and industry.

12. What are the factors influencing human population?

Unwanted fertility.

To increase the income for family and support Lack of knowledge on population control methods.

13. What is silicosis and asbestosis?

Silicosis is caused due to contamination of free silica. Asbestosis is due to asbestos fibres deposited in lungs.

14. List any four vector borne disease

Malaria, Dengue, Filaria, Encephalitis.

15. What is ameobiasis? Give its source.

This is a water borne disease, caused by amoeba i.e., Entameoba histolytica and charecterised by liquid stools with mucous and blood.

Source; food chain-fruits, vegetables, contaminated drinking water, cold drink, etc.

16. What are the two primary strains of HIV?

HIV-1 Strain.

HIV-2 strain.

17. List any two drugs used in treatment of AIDS.

AZT-Azidothymidine DDi-Dideoxyinosine.

18. List any two applications of information technology in environment.

Data on environmental degradation will help to restore the conditions Geographic information system by application of information technology will help in environmental conservation

Part B [16 Marks]

Unit I: Natural Resources

1. Explain Renewable & non-Renewable energy resources with examples.

2.

    a)   Compare nuclear power with coal. (6)

    b)   Write a note on energy conservation. (6)

    c)   What is Solar Space heating? Explain. (4)

3.

    a)   What are the causes of soil erosion & deforestation? Explain in detail. (8)

    b)   Discuss the consequences of Over utilisation of surface and Ground water. (8)

4. Write briefly on any four alternate source of energy.

5.

    a)   Write the effects of extracting and using mineral resources. (4)

    b)   Explain the scope & importance of Environment. (8)

    c)   Enumerate the benefits & draw backs of building of dams. (4)

6.

    a)   What are the effects of modern Agriculture practices. (8)

    b)   Discuss the causes of land degradation. (8)

Unit II: Ecosystems and Bio Diversity

1.
   a) What is an ecosystem? Describe the structure & function of various components of an ecosystem. (10)
   b) Explain the various threats to biodiversity. (6)

2.
   a) Define Ecological pyramids and explain different types of ecological pyramids. (6)
   b) Describe the types, characteristics features, structure& function of
      1). Forest ecosystem 2). Aquatic ecosystem. (10)

3.
   a) Explain In-Situ & Ex-Situ Conservation of Biodiversity. (8)
   b) What is meant by value of biodiversity? Explain different values of biodiversity.(8)

4. Write briefly on
   1. Ecological succession.
   2. Energy flow through an ecosystem.
   3. Hot spots of biodiversity. (4+4+8)

5.
   a) What is the biodiversity what are the reasons for decline for biodiversity. (10)
   b) Write short notes on: (1) Producers (2) Consumers (3) Decomposers. (3 X 2 = 6)

Unit III: Environmental Pollution

1.
   a) Discuss the major air pollutants and their impacts. (8)
   b) Explain the various methods of controlling air pollution. (8)

2.
   a) Discuss briefly the disposal of Municipal solid waste management. (8)
   b) Explain the causes, effects and control measures of water pollution. (8)

3.
   a) Discuss major air pollutants and their impacts. (6)
   b) What is Thermal pollution and explain their impacts?. (6)
   c) What are the sources of Radioactive pollution?. (4)

4.
   a) Explain in detail the role of individual in conservation of natural resources. (8)
   b) Explain the effects nuclear and Radiation pollution. (8)

5.

    a)   Write a brief note on of thermal pollution. (6)

    b)   Discuss the causes, effects & control measures of marine pollution. (10)

Unit IV: Social Issues and Green Approaches

1.

    a)   Explain forest conservation act. (8)

    b)   Write the factors influence the unsustainable to sustainable development. (8)

2.

    a)   Explain nuclear accidents and holocaust. (8)

    b)   Write notes on global warming. (8)

3.

    a)   Write a note on watershed management. (4)

    b)   Write briefly Bhopal disaster and Chernobyl disaster. (8)

    c)   Discuss briefly on environment protection act 1986. (4)

4.

    a)   Discuss the agenda for sustainable development. (8)

    b)   Write briefly on nuclear accidents. (8)

5. Write briefly on

    a)   Green House Effect. (4)

    b)   Acid Rain. (4)

    c)   Rain water Harvesting. (6)

    d)   Climate Change. (2)

Unit V: Human Population and Environment

1.

    a)   Explain the role of IT in environment and human health. (12)

    b)   Write short note on Value Education. (4)

2.

    a)   Write briefly on implementation of family planning programme. (6)

    b)   Write a note on AIDS in developing country. (6)

    c)   Discuss the factors influencing the family size. (6)

3.

    a)   Population explosion affects the environment seriously – discuss. (8)

    b)   Deterioration of environment leads to deterioration of human health – Justify. (8)

4.

a) Write a note on the following in relation to human population & environment.

A. Woman Child Welfare B. Human Rights C. Value Education. (8)

b) Write briefly on:

1. Population momentum.

2. Population Profile.

3. Age structure. (8)

5.

a) Discuss the growth of population can be successfully implemented. (8)

b) Discuss the factors influencing the family size. (4)

c) Discuss on the steps taken in India to prevent AIDS. (6)